SpringerBriefs in Energy

SpringerBriefs in Energy presents concise summaries of cutting-edge research and practical applications in all aspects of Energy. Featuring compact volumes of 50 to 125 pages, the series covers a range of content from professional to academic. Typical topics might include:

- A snapshot of a hot or emerging topic
- A contextual literature review
- A timely report of state-of-the art analytical techniques
- An in-depth case study
- A presentation of core concepts that students must understand in order to make independent contributions.

Briefs allow authors to present their ideas and readers to absorb them with minimal time investment.

Briefs will be published as part of Springer's eBook collection, with millions of users worldwide. In addition, Briefs will be available for individual print and electronic purchase. Briefs are characterized by fast, global electronic dissemination, standard publishing contracts, easy-to-use manuscript preparation and formatting guidelines, and expedited production schedules. We aim for publication 8–12 weeks after acceptance.

Both solicited and unsolicited manuscripts are considered for publication in this series. Briefs can also arise from the scale up of a planned chapter. Instead of simply contributing to an edited volume, the author gets an authored book with the space necessary to provide more data, fundamentals and background on the subject, methodology, future outlook, etc.

SpringerBriefs in Energy contains a distinct subseries focusing on Energy Analysis and edited by Charles Hall, State University of New York. Books for this subseries will emphasize quantitative accounting of energy use and availability, including the potential and limitations of new technologies in terms of energy returned on energy invested. The second distinct subseries connected to SpringerBriefs in Energy, entitled Computational Modeling of Energy Systems, is edited by Thomas Nagel, and Haibing Shao, Helmholtz Centre for Environmental Research - UFZ, Leipzig, Germany. This sub-series publishes titles focusing on the role that computer-aided engineering (CAE) plays in advancing various engineering sectors, particularly in the context of transforming energy systems towards renewable sources, decentralized landscapes, and smart grids.

All Springer brief titles should undergo standard single-blind peer-review to ensure high scientific quality by at least two experts in the field.

Zafar Khan Ghouri · Dilshad Hussain ·
David James Hughes

Hydrogen Production
from Seawater Electrolysis

Zafar Khan Ghouri
Net Zero Industry Innovation Centre
(NZIIC)
Teesside University
Middlesbrough, UK

Dilshad Hussain
HEJ Research Institute of Chemistry
University of Karachi
Karachi, Pakistan

David James Hughes
Centre for Sustainable Engineering
Teesside University
Middlesbrough, UK

ISSN 2191-5520 ISSN 2191-5539 (electronic)
SpringerBriefs in Energy
ISBN 978-3-031-73441-0 ISBN 978-3-031-73442-7 (eBook)
https://doi.org/10.1007/978-3-031-73442-7

This Springer imprint is published by the registered company Springer Nature Switzerland AG
The registered company address is: Gewerbestrasse 11, 6330 Cham, Switzerland

If disposing of this product, please recycle the paper.

Preface

A revolutionary research and development of green hydrogen generation through water electrolysis is gaining impetus worldwide as route to decarbonize our energy systems as around 40 % of worldwide Co2 emission originate from power generation. Further, water electrolysis using renewable source of electricity offers versatile approach to produce carbon free hydrogen. The shortage and necessity of high purified water for state-of-the-art electrolysis technology and nearly unlimited availability of seawater have led to considerable research efforts in developing direct seawater electrolysis technology for green H2 production. Groundbreaking innovative work on electrocatalysis for utilizing seawater directly as a feed has resulted in myriad of opportunities focused on advancing this state-of-the-art technology. In this perspective, this Springer Briefs in energy is intended to serve as a quick reference guide for students, researchers, academics, engineers, designers, stakeholders, and managers to offer an overview of the development of promising electrocatalysts for seawater electrolysis, whilst highlighting the pressing challenges.

Dr. Zafar Khan Ghouri
Vice Chancellor's Research Fellow
Net Zero Industry Innovation Centre
(NZIIC)
Teesside University
England, UK

Assistant Professor
International Centre for Chemical
and Biological Sciences
University of Karachi
Karachi, Pakistan

Contents

About the Authors

Dr. Zafar Khan Ghouri is a Vice Chancellor's Research Fellow in the Net-Zero Industry Innovation Centre at Teesside University and an Assistant Professor in the HEJ Research Institute of Chemistry, International Centre for Chemical and Biological Sciences (ICCBS) at the University of Karachi. Before joining TU, he was Research Scientist in the Department of Chemical Engineering at Texas A&M University at Qatar. Dr. Ghouri is endorsed by the Royal Academy of Engineering as a Global Talent, and he is currently an Academic Editor of PLOS ONE and Early Career Editorial Board member of Results in Engineering journals. His research interests can be divided into two focus areas, namely, 1) Synthesis of metallic nanoparticles decorated carbon nanostructures (nanofibers, nanotubes, or graphene) for neutral/alkaline seawater electrolysis and fuel cells, and 2) nanofabrication and characterization of novel pure organic (polymeric)/inorganic (ceramics) and composite materials for energy and environmental applications. He received his Ph.D. in engineering from the Department of BIN Fusion Engineering, Chonbuk National University under the prestigious Brain Korea (BK21) Fellowship. So far, he has published about 60 research papers in top-quality international peer-reviewed journals.

Dr. Dilshad Hussain works in the field of functional nanomaterials and their applications in textiles, sensing, membranes, and catalysis. His research work is focused on Metal-Organic Frameworks, Metal/Metal Oxide composites, Carbon Nanomaterials, Polymeric Materials, Quantum Dots, and Anisotropic Gold Nanoparticles. Dr. Dilshad's group research projects include multifunctional textiles, nano-functionalized membranes for environmental applications, and smartphone-assisted portable sensors. He has published more than 100 research papers in high-impact factor international journals with an accumulative Impact Factor above 500. He has also received two national research awards (Best Research Paper Award from HEC Pakistan and Research Productivity Award from Pakistan Council for Science and Technology Pakistan).

Prof. David James Hughes is a professor of low carbon materials and the Associate Dean for Research and Innovation within the Department of Engineering and has worked at the University since 2013. Prior to joining Teesside University David worked as a design and development technician in the polymer industry winning the Society of Polymer Engineers national award in 2008. David works closely with local trusts and industries. David is the Chair of the national IOM3 Polymer Group and sits on a number of government steering groups related to plastics. He coordinates the University Circular Economy and Recycling Innovation Centre (CERIC) with TWI.

Chapter 1
Introduction

Abstract In this chapter, fundamentals of water electrolysis with respect to kinetics and thermodynamics of water splitting reaction are overviewed. In addition, very briefly advantages and challenges of topic seawater electrolysis has been presented.

In recent years, hydrogen energy has attracted attention as an efficient and durable energy source. This interest stems from the fact that hydrogen energy is relatively stable. Electrolysis of water, i.e., the hydrogen evolution at the cathode and oxygen evolution at the anode, is one promising method to produce hydrogen. The thermodynamics of water electrolysis or water splitting is very straight forward due to the simplicity of the reaction in terms of reactant and products, although of the complexity of the reaction from an energetic point of view, as well as the equilibrium point for the reaction [1]. The electrochemical water splitting reaction can be simply present as:

$$\text{Cathodic hydrogen evolution reaction HER}: 2\,H_2O_{(l)} + 2\,e^- \rightarrow H_{2(g)} + 2\,OH^- \tag{1.1}$$

$$\text{Anodic oxygen evolution reaction OER}: 2\,OH^- \rightarrow {}^1\!/_2\,O_{2(g)} + H_2O_{(l)} + 2\,e^- \tag{1.2}$$

$$\text{Overall reaction water} - \text{splitting reaction}: H_2O_{(l)} \leftrightarrow H_{2(g)} + {}^1\!/_2\,O_{2(g)} \tag{1.3}$$

The energy required or released for this reaction can be determined thermodynamically from the Gibbs free energy according to the following equation:

$$\Delta_R G = \Delta_R H - T\Delta_R S \tag{1.4}$$

where ΔG_R is the Gibbs free energy of reaction, ΔH_R is the enthalpy of reaction, ΔS_R is the entropy of reaction, and T is the absolute temperature of the reaction. At normal temperature and pressure (NTP) of 25 °C and 1 bar, the value of $\Delta_R G^o$ is 237.2 kJ/mol, $\Delta_R H^o$ is 285.8 kJ/mol, and $\Delta_R S^o$ is 163.1 J/mol.K for liquid water

Z. K. Ghouri et al., *Hydrogen Production from Seawater Electrolysis*, SpringerBriefs in Energy, https://doi.org/10.1007/978-3-031-73442-7_1

Fig. 1.1 Cell voltage
dependence on temperature
for water electrolysis at
pressure of 1 bar. Reused
with the permission from ref
[5] Academic Press 2018

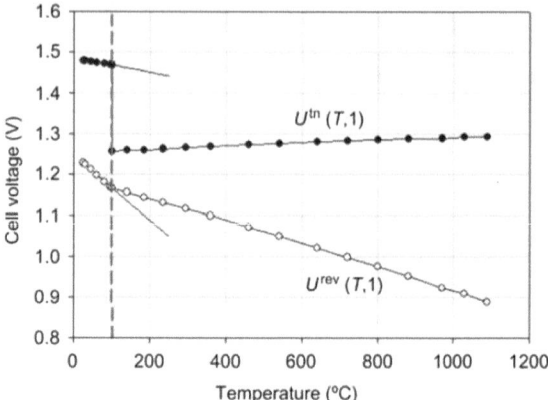

reactant and product hydrogen and oxygen in the gaseous phase. The positive value indicates that the reaction is non-spontaneous at NTP conditions and will require energy input [2–4]. The external electric potential or voltage required to drive the reversible electrochemical water splitting can be obtained from the Gibbs free energy as follow:

$$E_{rev} = -\Delta_R G / z.F \qquad (1.5)$$

where: z is the number of electrons involved in the reaction, and F is the Faraday constant of 96,485 C/mol. Accordingly, the standard cell voltage is -1.23 V, which is the theoretical value based on the reversible Gibbs free energy value [5]. However, to account for the process irreversibility and exegetic efficiency, i.e., thermal losses as $T\Delta_R S$, the actual electric potential or voltage is to be derived from the enthalpy value rather than the free energy as follow, leading to a potential of -1.48 V:

$$E_{irrev} = -\Delta_R H / z.F \qquad (1.6)$$

The dependence of water electrolysis reaction on temperature and pressure follows the collective dependence of $\Delta_R G$, $\Delta_R H$, and $\Delta_R S$, which are strong functions in temperature, as present in Fig. 1.1. Hence, the theoretical voltage required drops to -1.176, -0.978, and -0.920 V at 90, 800, and 1,000 °C, respectively [5].

Kinetics of chemical reaction, including electrochemical reactions determines the speed of reaction, or how fast the reaction will proceed Table 1.1. In water electrolysis, external electrical potential is applied to drive the electrolysis reactions at both cathode and anode, hence driving the overall reaction. The anodic OER and cathodic HER reactions are non-spontaneous, and in addition to the application of external energy, it requires the utilization of electrocatalyst i.e. cathode and anode electrodes to overcome the kinetic and thermodynamic barriers [6, 7]. In addition, anode electrocatalyst has to be highly active to boost the sluggish four-electron OER and be highly selective toward OER as well to minimize the competitive chlorine evolution

Table 1.1 Water electrolysis reactions in different media

Reaction	Acidic (pH=0)	Neutral (pH=7)	Alkaline (pH=14)
Anode	$2H_2O \rightarrow O_2 + 4H^+ + 4e^-$ $E_a = 1.23$ V	$2H_2O \rightarrow$ $O_2 + 4H^+ + 4e^-$ $E_a = 0.817$ V	$4OH^- \rightarrow$ $O_2 + 2H_2O + 4e^-$ $E_a = 0.404$ V
Cathode	$4H^+ + 4e^- \rightarrow 2H_2$ $E_c = 0.0$ V	$4H_2O + 4e^- \rightarrow$ $2H_2 + 4OH^-$ $E_c = -0.413$ V	$4H_2O + 4e^- \rightarrow$ $2H_2 + 4OH^-$ $E_c = -0.826$ V
Overall	$2H_2O \rightarrow O_2 + 2H_2 E_{cell} = +1.23$ V		

reaction (ClER), which is two-electron oxidation with only a single intermediate, when using seawater as feedwater [8, 9].

However, the development of green hydrogen production is quite limited due to production cost and unrestricted availability of potable water. Therefore, direct use of salt water as an electrolyte for hydrogen generation has become a hot research topic since potable water is a scarce commodity in many parts of the world, particularly Middle East and African Sahara region. Advantages and most common issues associated with sea water electrolysis is shown in Fig. 1.2.

a

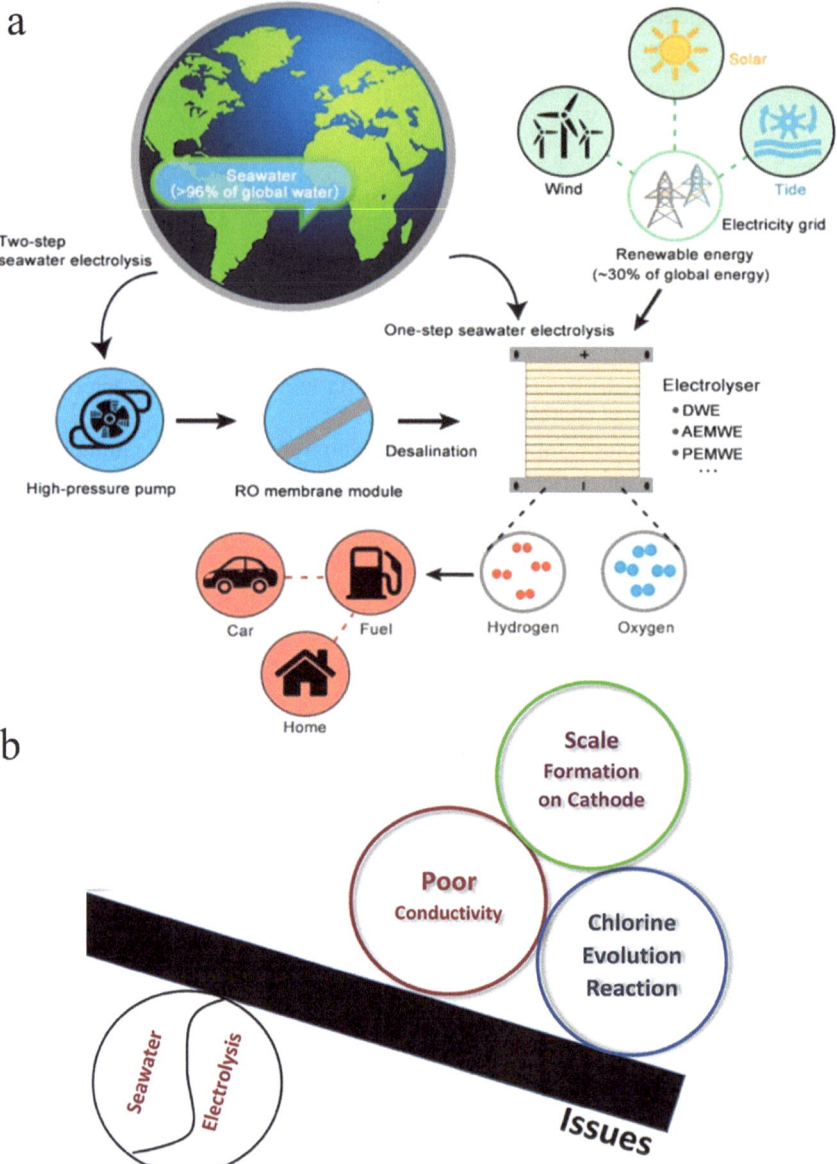

b

Fig. 1.2 (a) Different advantages of sea water electrolysis. Reused with permission from ref [10, 11] copyright 2022 Elsevier. (b) most common issues associated with seawater electrolysis reproduced with permission from ref [12] copyright 2022 Elsevier

References

1. Carmo M et al (2013) A comprehensive review on PEM water electrolysis. Int J Hydrogen Energy 38(12):4901–4934
2. Brauns J, Turek T (2020) Alkaline water electrolysis powered by renewable energy: a review. Processes 8(2):248
3. Millet P Grigoriev S (2013) Water electrolysis technologies. In: Brostow W (Ed) Renewable hydrogen technologies. production, purification, storage, applications and safety. Elsevier, Amsterdam
4. LeRoy RL, Bowen CT, LeRoy DJ (1980) The thermodynamics of aqueous water electrolysis. J Electrochem Soc 127(9):1954
5. Bessarabov D, Millet P (2018) PEM water electrolysis, Vol 1. Academic Press
6. Mohammed-Ibrahim J, Sun X (2019) Recent progress on earth abundant electrocatalysts for hydrogen evolution reaction (HER) in alkaline medium to achieve efficient water splitting–a review. J Energy Chem 34:111–160
7. Jamesh M-I, Sun X (2018) Recent progress on earth abundant electrocatalysts for oxygen evolution reaction (OER) in alkaline medium to achieve efficient water splitting–a review. J Power Sources 400:31–68
8. Dresp SR et al (2019) Direct electrolytic splitting of seawater: opportunities and challenges. ACS Energy Letters 4(4):933–942
9. Amikam G, Nativ P, Gendel Y (2018) Chlorine-free alkaline seawater electrolysis for hydrogen production. Int J Hydrogen Energy 43(13):6504–6514
10. Gao F-Y, Yu P-C, Gao M-R (2022) Seawater electrolysis technologies for green hydrogen production: challenges and opportunities. Curr Opin Chem Eng 36:100827
11. He W et al (2023) Materials design and system innovation for direct and indirect seawater electrolysis. ACS Nano 17(22):22227–22239
12. Asghari E et al (2022) Advances, opportunities, and challenges of hydrogen and oxygen production from seawater electrolysis: an electrocatalysis perspective. Curr Opin Electrochem 31:100879

Chapter 2
Principles and Mechanisms of Seawater Electrolysis

Abstract The electrocatalytic seawater splitting is an effective method for green hydrogen production to reduce carbon emission. Comparatively, seawater is more suitable than freshwater as a feed for the electrolysis thanks to its endless availability with challenges ahead. In this chapter, a brief introduction to the principle and reaction mechanism of seawater electrolysis is discussed, including the material design strategy for anode and cathode, as well as brief discussion on complete cell design has been presented.

Sea water electrolysis is a technique of splitting bonded elements by passing an electric current through them. An ionic compound is dissolved with a suitable solvent, like seawater, making the ions in the liquid available. Two inert metal electrodes or plates, such as platinum or iridium, are immersed in the seawater [1]. These plates are then connected to a DC electrical power source. Ions with the opposing charge are drawn to each electrode. Consequently, anions flow towards the anode, and cations move in the opposite direction. An electrical power source provides the power needed to separate the ions and induce them to amass at the appropriate surface. The ions at the probes absorb or release electrons, creating a concentration of the target element.

At 25 °C electrolysis, half reactions of pure water at 7 pH are.

Cathodic Reaction:

$$2H_2O + 2e^- \rightarrow H_{2(g)} + 2OH^- \quad E^\circ = -0.42V \tag{2.1}$$

Anodic Reaction:

$$2H_2O \rightarrow O_{2(g)} + 4H^+ + 4e^- \quad E^\circ = +0.82V \tag{2.2}$$

Overall this electrolysis is as follows.

$$2H_2O_{(l)} \rightarrow 2H_{2(g)} + O_{2(g)} \quad E^\circ = -1.24V \tag{2.3}$$

Pure water electrolysis has a negative cell potential which makes it thermodynamically unfavorable. An additional voltage (overvoltage) of approximately 0.6 V is required at each electrode due to the low ion concentration and the interfaces that need to be bridged by electrons. In practice, only a 2.4 V external voltage may be used to electrolyze clean water continuously. Scientists are looking for ways to make pure water electrolysis kinetically practical because it is thermodynamically impossible. One technique is to increase the number of particles existing by adding base, acid, or non-reacting salts, which increases conductivity. According to what is known about how saltwater splits, Chloride evolution reaction (CER) is more common when the pH is low, whereas OER and hypochlorite are both more likely to occur at the anode when the pH is high. Since seawater has a pH of 8.2, a bit alkaline, the main problem is solving the selective OER over hypochlorite production. Different membrane technologies have been made that allow only certain ions to pass through the membrane, keeping contaminants from getting to the cathode [2]. The anion exchange membrane alkaline water electrolyzes (AEMAWE) and photoelectrochemical-based water electrolyzes (PEMAWE) are two examples of alkaline water electrolysis membranes (AEMAWE).

Additionally, it is believed that the logical design of the HER electrode and using an alloy based HER electrode are the two most essential strategies to reduce the amount of corrosion on the electrode surface during selective HER. Adding a membrane to the electrolytes has several negative impacts, one of which is that it makes it more difficult for ions to move around. This, in turn, slows down the process by which ions flow through the water, reducing the device's efficiency [3].

Several practical technologies can produce huge volumes of hydrogen from seawater. The amount of hydrogen or energy needed to make 1 kg of hydrogen can be used to figure out how energy efficient the electrolyzers work. Before electrolysis in a Proton exchange membrane (PEM) electrolyzer, saltwater must be cleaned first with reverse osmosis (RO). Because this phase needs more energy, the plant's specific energy values may rise. Compared to the 47–66 kW h 1 kg of energy needed for the stack electrolyzer to treat fresh water, it has been shown that just 0.03 kW h of energy is needed for RO to treat 9 kg of saltwater [4], thought to be a small help. Suppose the single-step RO method before the electrolysis doesn't clean the water to the level needed for a PEM electrolyzer. In that case, more purification steps may be needed to keep the process going. This shows that the seawater electrolyzer uses a lot more energy now than it did before. In addition, a comprehensive feasibility analysis of directly electrolyzing seawater from offshore marine farms to make hydrogen finds that the procedure requires a lot of energy, making it less expensive in terms of energy and money.

As a result, one of the most crucial things to consider is the design of the electrolyzer reactor. Traditionally, seawater has been used in the symmetric reactor chamber, similar to an alkaline water electrolyzer (AWE), which uses a solution of 20–30% KOH in both compartments. In the case of the seawater electrolyzer, the choice of membrane is a problem. PEM doesn't matter because its electrolyte is acidic, which makes it harder for seawater to split [1].

The observed Gibbs free energy for the thermodynamically inclined chemical process of water electrolysis is approximately 237.2 kJ mol^{-1}. This process consists of two components: Oxygen evolution reaction (OER) at the anode and Hydrogen evolution reaction (HER) at the cathode. The two-electron transfer mechanisms used in seawater electrolysis provide molecular hydrogen and oxygen through a sequence of proton and electron coupling stages. The overpotential is the potential that is more than required for the total thermodynamic value for water splitting, which is 1.23 V [5].

Equations (2.1) to (2.4) illustrate the decomposition of water at various pH values. When pH levels are low:
In an acidic pH medium:

$$Anode : 2H_2O = O_2 + 4H + 4e^- \qquad (2.4)$$

$$Cathode : 2H^+ + 2e^- = H_2 \qquad (2.5)$$

In an alkaline pH medium:

$$Anode : 4OH- = O_2 + 2H_2O + 4e^- \qquad (2.6)$$

$$Cathode : C2H_2O + 2e^- = H_2 + 2OH^- \qquad (2.7)$$

In an alkaline medium, the OER kinetics is more favorable due to the presence of hydroxyl ions, but the HER process slows down [6]. Due to the interplay of several side reactions, the electrochemical state of seawater electrolysis will vary. A recent study by Strasser highlighted the possibility of redox reactions with seawater-dwelling species. The chloride appearance in NaCl is the main issue since it is the principal factor that interacts with, affects, and competes with the oxygen evolution process (OER) at the anode [6]. According to the chlorine Pourbaix diagram, the pH of the electrolyte will determine the reactions at the anode. The chlorine system in the case of 0.05 M sodium chloride aqueous solution with Pourbaix diagram for artificial seawater model and the advantages of using HSE over ASE as an energy saving tool and chlorine-free hydrogen production are depicted in Fig. 2.1.

The equilibrium between Cl and OCl is shown by the line (red) on the alkaline area of the equation, whereas the green line shows the equilibrium between Cl and Cl$_2$ on the acidic side of the equation. From a thermodynamic point of view, the green line shows that the H$_2$O and O$_2$ are stable. The blue region exhibits a constant potential difference of 480 mV for pH values above 7. On the basic side of the equation, the black and blue lines indicate where the Cl/HOCl and Cl$_2$/HOCl concentrations are in equilibrium, respectively. (b) A Pourbaix diagram that illustrates the over-potential ranges for OER as well as other chlorine oxidation processes at varying pH levels.

In acidic media

$$2Cl^- \rightarrow Cl_2 + 23^- E^\circ = +1.36 \text{ V} \qquad (2.8)$$

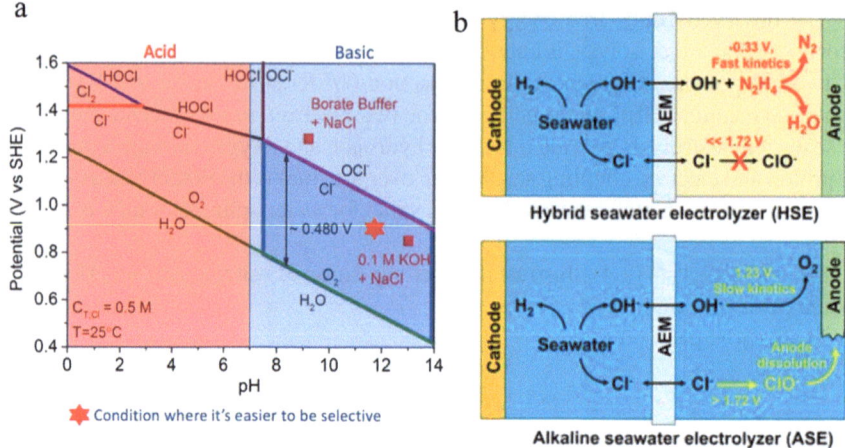

Fig. 2.1 (a) The chlorine system in the case of 0.05 M sodium chloride aqueous solution with Pourbaix diagram for artificial seawater model reused with permission from ref [7] copyright 2016 Wiley and (b) the advantages of using HSE over ASE as an energy saving tool and chlorine-free hydrogen production reused with permission from ref [8] copyright 2021 Nature

In alkaline media

$$Cl^- + 2OH^- \rightarrow ClO^- + H_2O + 2e^- \quad E^\circ = +0.89 \text{ V} \qquad (2.9)$$

The fact that the redox potential for conversion of bromide to bromine in seawater is relatively near to that of the redox process that occurs between chloride and chlorine (1331 V), which indicates that there is a high chance of interference between the two processes, saltwater is home to a plethora of different species. From a more practical point of view, the possibility that the promise it contains will never be realized is increased by the fact that its concentration in seawater is exceedingly low. Before comparing the redox reactions of CER and OER as a function of pH, the potential redox values of other species should first be determined as less interfering. This step is necessary before comparing the redox reactions of the CER and OER. In contrast to the conditions that must be present for the evolution of chlorine in acidic environments, the conditions that must be present for the development of the species known as hypochlorite must be present for chloride to go through the transition that leads to its production. Redox reactions always include the movement of two electrons and the production of a single intermediate, regardless of the circumstances. This final stage will be covered in greater detail in the following portion of the article. Because OER has a four-electron transfer, as demonstrated by Eqs. (2.1) and (2.3), Chloride evolution reaction (CER) likely has quicker kinetics than OER does. This is supported by the fact that CER does not have this transfer. The Pourbaix diagram indicates that OER is preferable to CER from a thermodynamic point of view. When the pH is alkaline, the difference in electrode potential between the CER and OER is

0.480 V. However, and this difference decreases when the pH is acidic. This seems to imply that in alkaline conditions, an over-potential value of less than 0.480 V is necessary to synthesize O_2 by OER while concurrently suppressing the synthesis of hypochlorite by CER because of the way that OER and CER interact. However, saltwater oxidation becomes problematic in an acidic environment, particularly at pH values below 3, due to the narrowing of the over-potential window to 180–350 mV, which is unfavorable for OER (b).

2.1 Material Design for Anode

In-depth research is done to create highly selective anodic electrodes for OER suppression of CER in seawater and other electrolyte solutions containing chloride [9]. Transition metal oxides and hydroxides have emerged as a promising class of electro catalysts for water oxidation reactions (OER) in alkaline water because of their effective catalytic activity established in these materials by defects and oxygen vacancies. The material's electrical structure, in some manner, influences these features. The electrical conductivity, available adsorption sites, or binding energies of the reaction decomposition products (M-OOH, M-OH) produced during the reaction determine the OER selectivity at the anode [10]. Anodic ways for separating seawater have been detailed in the sections that came before them; these approaches address the strong adversarial interaction between CER and OER. It is possible to increase electrodes' selectivity and, as a result, their capacity to change the electrical structure by doping them with heteroatom metals (also including Mn, Mo, Ni, Fe, or Co) or more active sites. Titanium's corrosion resistance to chloride and other contaminants and its identification by the chloro-alkali industries make it a popular choice as the inner substrate in a saltwater electrolysis anode. This section will focus on the advantages of using OER electrocatalysts over CER in seawater electrolysis and how to choose and construct such catalysts [11].

Bennett was the first to discover that a selective OER reaction in seawater could be catalyzed by an electrocatalyst made of MnOx in 1980. According to the study, MnO_2 is a more effective electrocatalyst for OER than CER. Because of its great stability in chlorine water, the substrate of TiO_2/RuO_2 has been widely employed as an anode in industrial chlorine production [12]. Porous MnO_2 coatings are electrochemically deposited on TiO_2/RuO_2 substrates to increase oxygen evolution productivity to 95% in saturated NaCl brine and over 99% in brine. The process of this work was greatly affected by the mass transport limitation of chlorine at the electrode. The diaphragm is MnO_2 coating, allowing the OER to act as the primary reaction at the anode without chlorine evolution. Trasatti reported in 1984 the catalytic activity of the oxides of several materials for CER and anodic OER and that his OER mechanism involved the hydroxides adsorption on metal oxides and an increment in valence states [13]. He reported that it could also be used for his CER. Though the metal oxide structure electronically determines the activity of the catalyst and kinetics, the chlorine selection is not dependent on the electrode material, by a linear slope for OER vs. cerium. After that, Hashimoto et al. added precious metals (Ru, Pt, Ir) and transition metals (Co, Ni, Fe, Sn, Mo) rare earth metals (La, Ce) to improve the electrode and confirmed the effect of OER in water. The electrodes MnO x are supported on a titanium (Ti) with an IrO_2 interlayer. The addition of up to 10 mol% of other metals improved the oxygen evolution efficiency, but studies in 0.1 M NaOH and 0.5 M NaCl electrolyte solutions at pH 8 showed that noble metals adversely affected the activity of His OER in chlorides, the activity of OER-containing solutions with different doping concentrations of noble metals. $MnOx/IrO_2/Ti$ electrodes with Mo as an additive showed the highest efficiency of oxygen evolution up to 90% at less (10 mol%) doping amount [14].

A single -Mn_2O_3 was formed at the low concentration of dopant, which was the cause of the OER activity being enhanced in this way. OER was unfavorable in a chloride-containing solution at the high concentration of Mo dopant due to the creation of a double-oxide composite and $MnMoO_4$. Hashimoto [15] continued his investigation into the OER activity in seawater by electrochemically anodizing a variety of mixed metal oxide coatings onto IrOx/Ti electrodes, including Mn-W, Mn-Mo, Mn-Mo-Sn, Mn-Mo-Fe, Mn-Mo-W, and Mn-Zn, during deposition of mixed metal oxides at the anode in each example, the IrOx served as a shielding layer to stop the Ti from oxidizing.

Based on these investigations, the inclusion of triple oxide chemistry with a structure of single-phase-$MnO2$ further improved the oxygen evolution efficiency, and the MnOx coating on the IrOx/Ti electrode suppressed the CER for selective oxygen evaluation. I have found that it is possible to create a seawater system. The highest oxygen evolution efficiency is obtained when MnOx contains the higher valence state Mo (VI) or W (VI). Additionally, by avoiding the oxidation of the inner Ti, which separates the electrocatalyst from the surface, this method addressed the stability of the electrodes. Different types of electrodes and their performance at different pH environment is shown in Fig. 2.2.

Theoretically validated electrocatalysts for efficient OER and CER are standard RuO_2 or IrO_2-based electrodes [21]. DFT analysis of the surface of $RuO_2(110)$ single crystals revealed the formation of peroxo groups that act as CER active sites, with Ocus atoms on neighboring Ru atoms used as intermediates for the surface. Therefore, when RuO_2 or IrO_2 is used, it isn't easy to achieve oxygen generation selectivity while suppressing chlorine generation, implying that doping was necessary to manipulate the electrical structure of the catalyst. To select the study OER for his CER in acid solutions containing chloride, Krtil, and his team used a freeze-drying method to produce his ruthenium oxide Zn-doped deposited on Ti mesh. LSV showed that pure RuO_2, as a suitable electrolyte for CER, resulted in Cl_2 evolution without O_2 at low pH, whereas RuO_2: Zn showed significant chlorine evolution under the same conditions. The process of selective OER was expected to avoid forming peroxo-bridge intermediates as a result of RuO_2 lattice rearrangement leading to defects in Zn insertion. With approximately 100% efficiency at 480 mV in a pH 13 brine artificial electrolyte, the NiFe-LDH electro-catalyst created by Strasser and his colleagues utilizing a solvothermal method is a noteworthy option for OER [7]. The electrocatalytic activity and selectivity of chlorine-containing electrolytes are

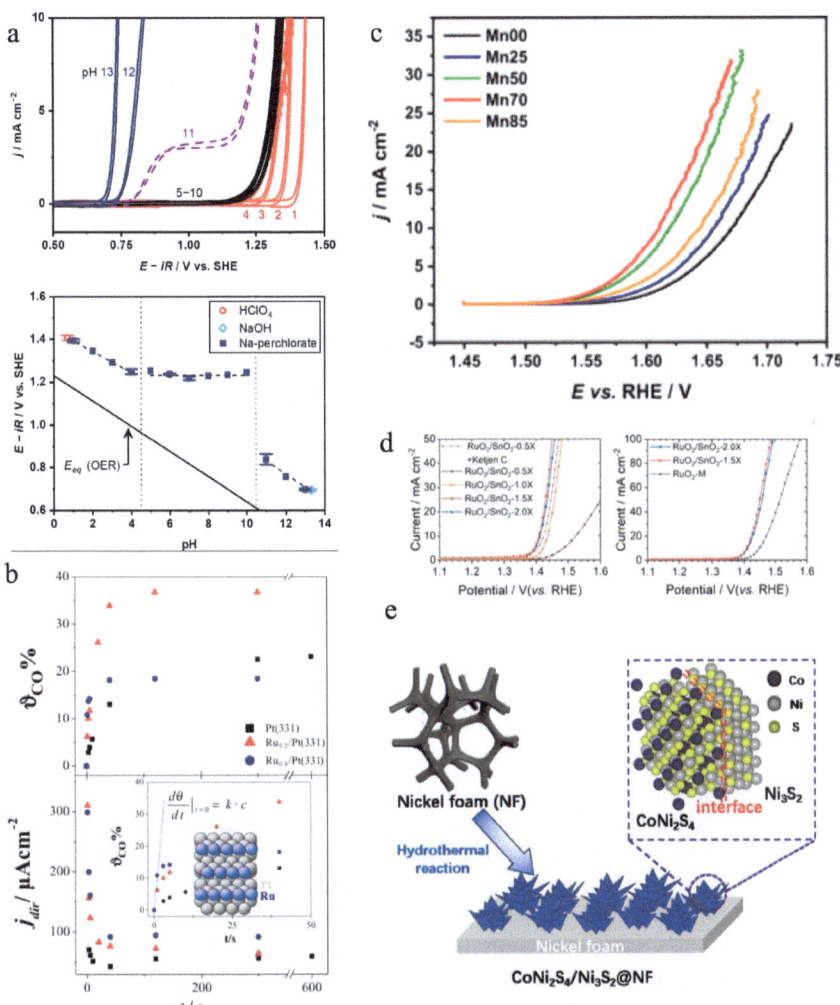

Fig. 2.2 (a) Oxygen evolution reaction (OER) of IrO at different acidic and basic pH reused with permission from ref [16] copyright 2020 Elsevier (b) the record of CV and MSCV of $m/z = 44$ and effect of Ru-step decoration on the relative charge coverage of methanol adsorbate reused with permission from ref [17] copyright 2015 Elsevier (c) Linear sweep voltammetry of Ir-Mn films recorded in 0.1 M sulfuric acid reused with permission from ref [18] copyright 2021 Elsevier (d) The OER in 0.1 M HCLO$_4$ for LSV curves of RuO$_2$/SnO$_2$-xX reused with permission from ref [19] copyright 2023 Wiley and (e) illustration of preparation of COnI2S4/Ni3S2@NF electrode reused with permission from ref [20] copyright 2020 Elsevier

limited to near-neutral pH, and under these conditions, he shows that CER suppression is difficult. Similarly, Chen et al. have prepared low-cost layered structures of CoFe-LDH nanomaterials loaded onto Ti substrates as electrodes to study OER-selective electrocatalysis to decompose seawater without the use of buffers. CoFe-LDH nanoparticles exhibited improved catalytic activity and reasonable O_2 Faradaic efficiency at an overpotential of 530 mV in pH 8 electrolyte due to the collegial effect of cations of different metals [22]. Kuang et al. used an amorphous layer of NiFe hydroxide deposited by placing electrodes on the surface of a NiSx sulfide-coated Ni substrate to create a corrosion-resistant anode for seawater fission [23]. The upper most amorphous NiFe hydroxide layer acts as an efficient electrocatalyst for OER, while the bottom NiSx layer not only improves the electrical conductivity of the electrode but also provides a cation-selective layer that is resistant to corrosion by chlorine which also produces polyatomic, anion-rich electrodes in seawater. It can also operate at the high current densities typically required in the industry. Recently, porous S-doped NiFe oxyhydroxide prepared by using a scalable ultrafast technique at ambient temperature and effective for OER in seawater, with moderate over voltages achieved very high current densities.

2.2 Material Design for Cathode

The noble metal Pt is known to have a low kinetic energy barrier to dissociating water molecules, 0.89 eV on the Pt (111) surface [24]. This lower kinetic energy barrier makes Pt a conventional electrocatalyst for HER. The situation for saltwater, however, is exceedingly difficult because of contaminants, which could impede the action and result in catalyst poisoning. Seawater splitting must find an alternative method to establish a corrosion-free but effective cathode. Pt-based alloys, including D-block metals, might be considered a replacement for pure Pt because they exhibit promising electrocatalytic activity and improved stability toward the HER mechanism. It is thought that the competitive dissolving rate of reaction in alloys contributes towards the electrode's improved anti-corrosion capabilities. The Brewer-Engel coupling theory, which defines the potential d-orbital overlap of metals in intermetallic compounds and predicts a stable system of kinetically favorable HER mechanisms, is useful for designing active alloy-based HER catalysts. This hypothesis served as a guide for developing several Pt-based alloys for alkaline water electrolysis. This method turned out to be more effective in seawater decomposition, as reported by Li et al.. By incorporating Ru and a 3d transition metal (M) as a guest metal into FCC platinum (Pt) as the host metal without any visible change in the lattice structure of the host metal, they formed a Pt-Ru-M alloy (M = Fe, Mo) was created., Co and Ni Cr), using the electrode placement method on a Ti mesh [25].

Lattice aberrations induced by the introduction of guest metals provide useful active sites for enhancing H+ adsorption. Compared with other transition metals, Mo exhibited the highest Mo–H bond strength in Pt–Mo alloys, resulting in the best catalytic performance Importantly, the preparation of this alloy showed coveted stability over 172 h in seawater electrolysis due to the excellent d-band interaction of the PtMo alloy, according to Brewer's theory, and the increased surface area of the alloy. Subsequently, Zheng et al. reported using a similar methodology to demonstrate the long-term stability of Pt-M alloys over 170 h at the HER of seawater electrolysis [26]. Thermodynamic analyzes of the corrosion behavior of nickel and molybdenum in chlorine-containing environments evaluated by Galetz and his colleagues provided strong support for such studies [27]. The extraordinary durability of his Pt–Mo alloy materials in chlorinated brines may be attributed to the fact that Mo reacts with chlorine much more slowly than Ni in hypoxic environments. Reaction order of the PtN_4 catalyst and Ferro-De Battisti test plots for the chlorine evolution reaction (CER) over the PtN_4 catalyst is shown in Fig. 2.3.

Even if Pt-containing alloys are produced with good stability, and the cost of cathode materials is reduced to some extent, the cost of electrodes is still high and cannot be used for increasing hydrogen production from seawater. Several research groups have developed materials with high catalytic activity and non-corrosiveness in chlorine-containing electrolytes as alternatives to expensive Pt-based cathodes used in seawater splitting. Covering the effective electrocatalyst with a protective carbon coating is a practical way to prevent seawater corrosion of the electrocatalyst during water splitting. To efficiently decompose seawater, Gao et al. [30] have fabricated co-embedded N-doped carbon nanotubes tested at different pH settings. This material showed good stability at three different pH values, including buffered seawater and untreated seawater, but the catalytic activity might not have been as strong as pure Pt as an electrode. By using the pyrolysis technique, Ma et al. have prepared CoMoP nanocrystals with several layers of N-doped carbon shells [31]. The electrocatalyst promoted their HER activity over a wide pH range and showed Faradaic efficiencies up to 92.5% in real saline water. This result suggested that the carbon coating of CoMoP protected her CoMoP from saltwater contaminants, improved the free energy of H adsorption, and enhanced HER activity by enhancing water adsorption. Another method Lu and his colleagues used was observing the deposition of Na^+, Ca^{2+}, or Mg^{2+} salts on the surface of his Mn-doped Ni/NiO on the Ni foams they prepared, which revealed the activity-blocking parts and affecting performance [32]. They recommended a moderate acid treatment to restore the electrocatalyst's functionality.

Molybdenum disulfide, with its active S-sites uncovered, is thought to be a very strong catalyst for the HER mechanism due to its 2D layered structure, which makes the active sites more accessible for electrolytes. Seawater splitting experiments have also been performed, with MoS_2 identified as an effective electrocatalyst for HER. Using the hydrothermal technique, Miao and colleagues created hierarchical Ni-Mo-S nanosheets on carbon cloth [33]. According to their study, adding Ni atoms to MoS_2 led to defects in the active catalytic sites, affecting the multilayer MoS_2's electronic structure. Due to its mesoporous structure and more exposed active sites, the material with the ideal 1:1 Ni: Mo ratio showed excellent catalytic activity for HER in neutral buffer electrolytes and real saline.

Chen and his colleagues have developed a porous 3D MoS_2 quantum dot aerogel suitable as a cathode for seawater fission [34]. A large surface area was found in the airgel. It was an excellent electrocatalyst, performing similarly to conventional Pt in

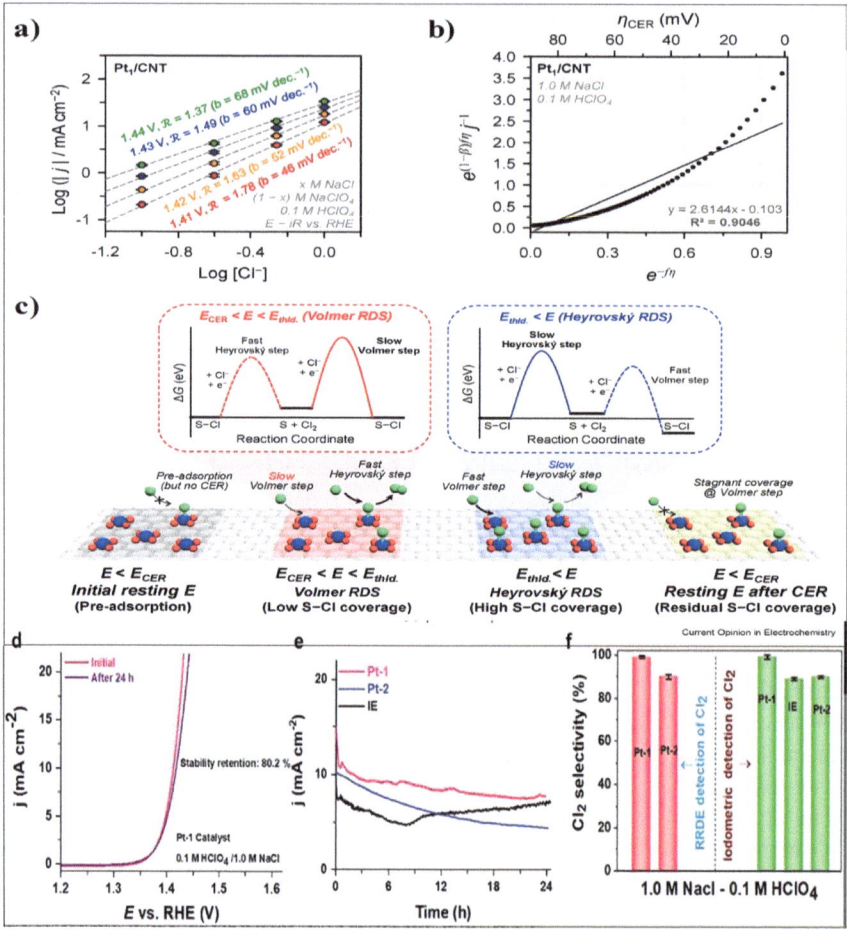

Fig. 2.3 (a) Reaction order of the PtN$_4$ catalyst calculated by log *j versus* log [Cl$^-$] plots. (b) Ferro-De Battisti test plots for the chlorine evolution reaction (CER) over the PtN$_4$ catalyst. (c) Reverse Volmer–Heyrovsky mechanism. Reused with permission from ref. [28], Copyright 2021, American Chemical Society. (d) ClER polarization curves before and after durability tests. (e) Current density changes after durability assessments (applied potential = 1.42 V). (f) ClER selectivity obtained by detecting Cl$_2$ using rotating ring-disk electrode (RRDE) and iodometric titration. Reused with permission from ref. [29], Copyright 2023 Wiley

a wide pH range and seawater, with excellent stability over 10,000 cycles. Like metal phosphides and sulfides, metal nitrides offer good prospects as electrocatalysts for HER due to their unique electronic structure and electrical conductivity. However, the stability problem remains difficult to solve due to the low valence of the metal. The problem of N enrichment in metal nitrides can be overcome by increasing the valence state of metal atoms to increase corrosion resistance.

Mo$_5$N$_6$ nanosheets, an N-enriched metal nitride, were created by Jin et al. using the Ni-induced salt template approach [35]. Their study showed that, in addition to increasing the surface area, the incorporation of N into MoN modifies the d-band center closest to the Pt metal, resulting in improved HER catalytic activity, as good as Pt, and good stability. (100 h). Yu et al. have fabricated effective nickel phosphide (NixP)-based electrodes sandwiched between NiCoN nanostructures for HER in untreated brine [36]. At greater current densities and with strong stability, the electrocatalyst displayed a comparable performance to Pt's benchmarking performance. It was thought that NixP and NiCoN's corrosion resistance characteristics kept the electrocatalyst stable and maintained performance in untreated seawater. Xiu and colleagues developed hollow low Pt catalysts tailored for MXene with a wide pH range for seawater to achieve the typical 20 wt% Pt catalytic activity of HER [37]. Due to the increased surface area and conductivity, the low Pt hollow foam exhibited greater catalytic activity for HER in seawater. In addition to natural seawater, Wu and his colleagues developed his 2D CoxMo2xC/MXene/NC electrocatalysts with low overpotential and fast HER kinetics. It was believed that careful interfacial engineering was required to achieve the such high catalytic activity, excellent durability, and Faradaic efficiency in natural salt water. In recent studies, a mixture of metal sulfides and metal carbides, known as VS2@V2C, also showed higher catalytic activity than conventional Pt in both natural brines and all pH ranges [38]. The tight wrapping of V2C by VS2 was thought to be responsible for this high activity, resulting in reduced free energy for hydrogen adsorption and excellent electrocatalytic activity.

2.3 Complete Cell Design

Constructing highly active and stable dual-functional electrocatalysts for anodic oxygen evolution and cathodic hydrogen evolution is challenging but very important. Various dual functional catalysts, such as nitrides and metallic chalcogenides, have been reported for alkaline water splitting, with widely varying electrical properties and morphologies. Considering the stability and selectivity of the catalyst, there are not many reports on the direct decomposition of seawater. Using his one-step calcination to control the Co to Se mass ratio, Zhao and his colleagues fabricated a series of cobalt selenide electrodes. In this method, low Co charge states promote HER activity, and high Co charge states favor OER, resulting in global seawater decomposition reaching current densities of Helium 10.3 mA cm2 at 1.8 V [39]. To dissociate water molecules, Zhao et al. have fabricated NiNS electrodes with interfaces between Ni$_3$S$_2$ and Ni$_3$N and frequently open electrochemical areas. Dissociative water adsorption with such a material design and a penetrating surface is favorable, resulting in total seawater decomposition reaching current densities of 48.3 mA cm2 at 1.8 V [40]. The transition metal hexacyanometallate (MHCM-z-BCC) with a basic cobalt carbonate (BCC) as a conductive core formed on a pretreatment carbon cloth was used by Hsu et al. to develop a PV-driven seawater splitting device. NiMoS was used as the cathode [41]. NiMoS was used as the cathode. Even after 100 h, he shows no signs of Cl2 evolution, proving that the electrode combination is very successful in neutral seawater fission. Scientists argued that BCC selectivity, with MHCM-z greater than OER over CER, was an important factor in efficiently partitioning seawater into O_2 and H_2.

References

1. Khan M et al (2021) Seawater electrolysis for hydrogen production: a solution looking for a problem? Energy Environ Sci 14(9):4831–4839
2. Chang J et al (2021) Dual-doping and synergism toward high-performance seawater electrolysis. Adv Mater 33(33):2101425
3. Park YS et al (2021) High-performance anion exchange membrane alkaline seawater electrolysis. Journal of Materials Chemistry A 9(15):9586–9592
4. Meier K (2014) Hydrogen production with sea water electrolysis using Norwegian offshore wind energy potentials. Int J Energy Environ Eng 5(2):1–12
5. Baniasadi E (2012) Development of a new hybrid photochemical/electrocatalytic water splitting reactor for hydrogen production: design, analysis and experiments
6. Zhang F et al (2021) Rational design of oxygen evolution reaction catalysts for seawater electrolysis. Trends in Chemistry 3(6):485–498
7. Dionigi F et al (2016) Design criteria, operating conditions, and nickel–iron hydroxide catalyst materials for selective seawater electrolysis. Chemsuschem 9(9):962–972
8. Sun F et al (2021) Energy-saving hydrogen production by chlorine-free hybrid seawater splitting coupling hydrazine degradation. Nat Commun 12(1):4182
9. Yao Y, Gao X, Meng X (2021) Recent advances on electrocatalytic and photocatalytic seawater splitting for hydrogen evolution. Int J Hydrogen Energy 46(13):9087–9100
10. Wang XH et al (2021) Doping modification, defects construction, and surface engineering: Design of cost-effective high-performance electrocatalysts and their application in alkaline seawater splitting. Nano Energy 87:106160
11. Li R et al (2022) Synergistic interface engineering and structural optimization of non-noble metal telluride-nitride electrocatalysts for sustainably overall seawater electrolysis. Appl Catal B 318:121834
12. Feng S et al (2022) Recent progress in seawater electrolysis for hydrogen evolution by transition metal phosphides. Catal Commun 162:106382
13. Trasatti S (1984) Electrocatalysis in the anodic evolution of oxygen and chlorine. Electrochim Acta 29(11):1503–1512
14. Bennett J (1980) Electrodes for generation of hydrogen and oxygen from seawater. Int J Hydrogen Energy 5(4):401–408
15. E-M AA, Kumagai N, Hashimoto K (2009) Mn-Mo-W oxide anodes for oxygen evolution in seawater electrolysis for hydrogen production. Mater Trans 50(8):1969–1977
16. Nishimoto T et al (2020) Microkinetic assessment of electrocatalytic oxygen evolution reaction over iridium oxide in unbuffered conditions. J Catal 391:435–445
17. Abd-El-Latif A et al (2015) Insights into electrochemical reactions by differential electrochemical mass spectrometry. TrAC, Trends Anal Chem 70:4–13
18. Pascuzzi MEC, Hofmann JP, Hensen EJ (2021) Promoting oxygen evolution of IrO2 in acid electrolyte by Mn. Electrochim Acta 366:137448
19. Huang B et al (2023) Dense-Packed RuO2 nanorods with in situ generated metal vacancies loaded on SnO2 nanocubes for proton exchange membrane water electrolyzer with ultra-low noble metal loading. Small 19(34):2301516
20. Dai W et al (2020) Flower-like CoNi2S4/Ni3S2 nanosheet clusters on nickel foam as bifunctional electrocatalyst for overall water splitting. J Alloy Compd 844:156252

21. Exner KS, Sohrabnejad-Eskan I, Over H (2018) A universal approach to determine the free energy diagram of an electrocatalytic reaction. ACS Catal 8(3):1864–1879
22. Cheng F et al (2017) Synergistic action of Co-Fe layered double hydroxide electrocatalyst and multiple ions of sea salt for efficient seawater oxidation at near-neutral pH. Electrochim Acta 251:336–343
23. Yu L et al (2020) Ultrafast room-temperature synthesis of porous S-doped Ni/Fe (oxy) hydroxide electrodes for oxygen evolution catalysis in seawater splitting. Energy Environ Sci 13(10):3439–3446
24. Gao Q et al (2019) Structural design and electronic modulation of transition-metal-carbide electrocatalysts toward efficient hydrogen evolution. Adv Mater 31(2):1802880
25. Li H et al (2016) Robust electrocatalysts from an alloyed Pt-Ru-M (M= Cr, Fe Co, Ni, Mo)-decorated Ti mesh for hydrogen evolution by seawater splitting. Journal of Materials Chemistry A 4(17):6513–6520
26. Zheng J et al (2018) Seawater splitting for hydrogen evolution by robust electrocatalysts from secondary M (M= Cr, Fe Co, Ni, Mo) incorporated Pt. RSC Adv 8(17):9423–9429
27. Galetz M, Rammer B, Schütze M (2015) Refractory metals and nickel in high temperature chlorine-containing environments-thermodynamic prediction of volatile corrosion products and surface reaction mechanisms: a review. Mater Corros 66(11):1206–1214
28. Lim T et al (2021) General efficacy of atomically dispersed Pt catalysts for the chlorine evolution reaction: potential-dependent switching of the kinetics and mechanism. ACS Catalysis 11(19):12232–12246
29. Ha M et al (2023) High-Performing atomic electrocatalyst for chlorine evolution reaction. Small 19(20):2300240
30. Gao S et al (2015) Electrocatalytic H 2 production from seawater over Co, N-codoped nanocarbons. Nanoscale 7(6):2306–2316
31. Ma Y-Y et al (2017) Highly efficient hydrogen evolution from seawater by a low-cost and stable CoMoP@ C electrocatalyst superior to Pt/C. Energy Environ Sci 10(3):788–798
32. Lu X et al (2018) A sea-change: manganese doped nickel/nickel oxide electrocatalysts for hydrogen generation from seawater. Energy Environ Sci 11(7):1898–1910
33. Miao J et al (2015) Hierarchical Ni-Mo-S nanosheets on carbon fiber cloth: a flexible electrode for efficient hydrogen generation in neutral electrolyte. Sci Adv 1(7):e1500259
34. Wang S et al (2015) A new molybdenum nitride catalyst with rhombohedral MoS_2 structure for hydrogenation applications. J Am Chem Soc 137(14):4815–4822
35. Jin H et al (2018) Single-crystal nitrogen-rich two-dimensional Mo_5N_6 nanosheets for efficient and stable seawater splitting. ACS Nano 12(12):12761–12769
36. Yu L et al (2020) Hydrogen generation from seawater electrolysis over a sandwich-like NiCoN| Ni x P| NiCoN microsheet array catalyst. ACS Energy Lett 5(8):2681–2689
37. Xiu L et al (2020) Multilevel hollow MXene tailored low-Pt catalyst for efficient hydrogen evolution in full-pH range and seawater. Adv Func Mater 30(47):1910028
38. Wu X et al (2019) Engineering multifunctional collaborative catalytic interface enabling efficient hydrogen evolution in all pH range and seawater. Adv Energy Mater 9(34):1901333
39. Zhao Y et al (2018) Charge state manipulation of cobalt selenide catalyst for overall seawater electrolysis. Adv Energy Mater 8(29):1801926
40. Zhao Y et al (2019) Interfacial nickel nitride/sulfide as a bifunctional electrode for highly efficient overall water/seawater electrolysis. Journal of Materials Chemistry A 7(14):8117–8121
41. Hsu SH et al (2018) An earth-abundant catalyst-based seawater photoelectrolysis system with 17.9% solar-to-hydrogen efficiency. Adv Mater 30(18):1707261

Chapter 3
Electrocatalyst Design for Hydrogen Evolution Reaction

Abstract Development of highly efficient, stable, and cost effective electrocatalysts to meet the needs of highly effective green hydrogen production from seawater splitting is still challenging. In this perspective, some advances electrocatalysts with superior performance, such as noble metal based electrocatalysts, carbon incorporated noble metals based electrocatalysts, metal phosphides based electrocatalysts, metal oxides and hydroxides based electrocatalysts, transition metal carbides (TMCs) based electrocatalysts, transition metal phosphides based electrocatalysts, transition metal chalcogenides based electrocatalysts, metal nitrides based electrocatalysts, and etc., have been synthesized for hydrogen production from the seawater splitting. In order to gain insight into the current advancement on electrocatalysts for seawater hydrogen evolution reaction, this chapter briefly epitomize the general design criteria for electrocatalysts in seawater hydrogen evolution reaction in recent years.

In the electrochemical process of splitting water, the cathodic process is called the Hydrogen evolution reaction (HER, $2H^+ + 2e^- = H_2$). The hydrogen evolution process (HER) is a prototypical two-electron transfer reaction with a single catalytic intermediate. It has the potential to produce hydrogen, which is both an essential reagent in chemistry and a useful source of fuel. Using renewable energy sources in the operation of the HER might provide a renewable reservoir of hydrogen fuel that could be used, stored, and transported to an emission-free combustion engine or fuel cell. To achieve high energy efficiency in water splitting, a catalyst is necessary to lower the overpotential needed to drive the HER. Platinum is the most widely used HER catalyst because it only needs small overpotentials, even at high reaction rates in acidic environments. Unfortunately, Pt's high price and small availability limit its application in advanced technology.

The catalytic activity of the Hydrogen evolution reaction (HER) was investigated to establish the basic parameters that govern this activity. It has improved its comprehension of the surface structures and features that control HER activity and stability by developing new catalyst materials for the HER. The work has created numerous earth-abundant HER catalysts, particularly sulfide- and phosphide-based materials, with activities that resemble those seen for platinum by using this understanding in next-generation catalyst design. The literature frequently reports on novel hydrogen evolution reaction (HER) catalysts without considering the mass-transport requirements, intrinsic activity, or applied catalyst loading. By improving mass loading and the associated active surface area, high geometric current densities have been worked hard to achieve.

However, while the geometric current density is significant from an applied standpoint, it doesn't accurately represent the inherent catalytic activity that results from adjusting the catalyst's electronic structure. The only measurable sign of intrinsic activity is the turnover frequency (TOF), which is the rate at which new molecules (such as H2) are generated per second per site.

Additionally, certain typical errors reduce the accuracy and value of the reported results in the literature, such as subpar counter electrodes, HER reaction performed without hydrogen saturation, and comparisons to substandard measurements.

Catalysts for hydrogen evolution reaction (HER) can be categorized into two groups: based and non-noble metals. Numerous methods are being explored to improve HER performance and reduce the price of the noble-metal-based electrocatalyst. The electrocatalytic performance and kinetic characteristics of these catalysts under various reaction circumstances are compared [1]. For example, when Pt is alloyed with less expensive transition metals, its effectiveness may rise, and the alloy's synergistic effects may alter the environment of the electronic system to increase activity. When combined with other water dissociation promoters, platinum can significantly increase alkaline HER activity, which is essential for its application in real industrial settings.

HER electrocatalysts based on non-noble metals have gathered a lot of attention because they are both relatively affordable and readily available on Earth. Electrocatalytic HER has significantly advanced thanks to electrocatalysts of non-noble metal. These are some examples of transition metal chalcogenides, transition metal carbides, and transition metal phosphides [2].

The Hydrogen Evolution Reaction (HER) is a two-electron transfer mechanism that is the major half-reaction in water electrolysis that produces hydrogen at the cathode. The environment has a significant influence on how this HER works. Three potential reaction stages are available for the HER reaction in acidic environments.

$$H^+ + e^- = \text{Had} \tag{1a}$$

$$H^+ + e^- + \text{Had} = H_2 \tag{1b}$$

$$2\text{Had} = H_2 \tag{1c}$$

The Volmer step (1a) is the first step in synthesizing adsorbed hydrogen. The H_2 can then be produced via the hydrogen evolution reaction, which can be carried out via the Heyrovsky step (1b), the Tafel step (1c), or both. About the HER's reaction in basic buffers [3]. The Volmer step (2a) and the Heyrovsky step (2b) are two possible reactions based on the given equations [4].

Table 3.1 Electrocatalysts with their reported HER performance

Catalyst	Electrodes	Overpotential [V]	Tafel Slope [mV/dec]	Current Density	Ref
NiRuIr_G	DRP-110	0.08 V	48	40	[6]
Ni/NiMoN	Pt	37 mV	51	10	[7]
Ni–Fe–Mo	Pt mesh	270	40	0.5	[8]
Ni–Fe–Co	Pt mesh	265	37	0.5	[8]
NiS2@GO	Ag/AgCl	57	47	10	[9]
((Co,Fe)PO4)	Graphite rod	137	-71	10	[10]
(Ni-MoO2)	GCE	234	181	10	[11]
CoNiP/CoP	NF	36	70	10	[12]
Cu2S@Ni	Cooper foam	200	95.1	500	[13]
CoP/Co2P	GC	454	104	10	[14]
FeP@CoP	Co/NF	50	50.1	10	[15]
Ni-MoN	Copper foam	61	35.5	100	[16]
NiMo@C3N5	GC	80	68.3	10	[17]
NiCoHPi@Ni3N/NF	Ni3N/NF	174	79.8	100	[18]
Co6W6C	Ag/AgCl	50	40.99	10	[19]
MoNi/NiMoO4	NF	256	46.3	10	[20]
Pt@CoMo2S4-NGNF	PC@CMS-NG	27	32	10	[21]
Ni/MoO2@CoFeOx	NF	39	65	10	[22]
Rh-WO3	GC	48	131	10	[23]
Ru, W–NiSe2/NF	NF	100	119.7	10	[24]
NiMn/Ti	Graphite rod	79.3	214	10	[25]
Co/W5N4	NF	30	91.8	10	[26]

$$H_2O + e^- = OH^- + Had \tag{2a}$$

$$H_2O + e^- + Had = OH^- + H_2 \tag{2b}$$

It is essential to find a middle ground between hydroxy adsorption (OHad), hydrogen adsorption (Had), and water dissociation for HER to function properly in alkaline settings. Theoretical models suggest a connection between HER activity and hydrogen adsorption (Had) [5]. It is generally believed that the free energy of hydrogen adsorption, abbreviated as GH, can be used to characterize a material that participates in hydrogen evolution. For the HER process to work properly, there should be a satisfactory amount of hydrogen binding energy [4]. The list of catalysts for HER is summarized in Table 3.1.

3.1 Noble Metal Based Electrocatalysts

The HER catalytic activity of noble metals, like the Pt group metals (PGMs, which include Rh, Ir, Ru, Pd, and Pt), is exceptional. Pt is the top-ranked point on the volcanic curve [27]. The broad commercial application of catalysts centered on noble metals is improbable due to the prohibitively high cost of the catalysts and their limited capacity for storage, despite these catalysts being very effective. To address this challenge, noble metal catalysts are used with alloys with a low metal loading [28].

By alloying with transition metals, Pt can be used far more effectively. In addition to increasing HER electroactivity, the alloys' synergistic effects could cause a shift in the surrounding electrical field [29]. Ultralow-loading Pt content (7.7%) Ni nanosheets coated with PtNi nanoparticles and arrayed on carbon cloth (PtNi-Ni NA/CC) were seen to grow in situ by Sun et al. [30]. The HER activity of these Ni nanosheets is significantly higher than that of Pt/C (20%) when tested in 0.1 M KOH at a current density of 10 mA cm2 and a restricted overpotential of 38 mV. Significantly, a 90-h test of catalytic activity reveals that it is also long-lasting. Because of the downshift in the d-band center of Pt, the HER performance of PtNi-Ni NA/CC is significantly improved since the adsorption energy of oxygenated species (OH*) on the surface of Pt atoms is decreased [31].

The activity of HER, a standard by which Pt electrocatalyst is measured, is often lower in the alkaline medium than in acidic media. Because there is inadequate water dissociation on the surface of the platinum, HER activity is considered ineffective. Increasing alkaline HER activity is a prominent goal, and the most common method for reaching this goal now involves combining Pt with water dissociation promoters. The capacity to modify the surface metal composition of the many Pt-based electrocatalysts for HER is crucial for enhancing their electrocatalytic performance. Recent research by Markovic and colleagues has shown that nanosized $Ni(OH)_2$ clusters may be synthesized in a controlled manner on Pt electrode surfaces [32]. Compared to the Pt that was being used at the time, they discovered that this approach enhanced HER activity by 8. The Ni $(OH)_2$ cluster's edges on the surface of Pt promote water dissociation and the formation of M-Had intermediates. When the intermediates of adsorbed hydrogen combine, H_2 is produced.

Surface engineering can be used to produce PtNi-O nanoparticles that have a better NiO/PtNi interface. This was demonstrated by Huang et al., who used $Ni(OH)_2$ and Pt to operate well together. This interface structure converts to $Ni(OH)_2$ in basic conditions, resulting in the formation of a chemically comparable surface to $Ni(OH)_2$/Pt. The catalyst exhibited a moderately low HER overpotential of 39.8 mV when subjected to a current density of 10 mA cm^2 with only 5.1 g cm^2 of Pt being provided [33].

After undergoing the annealing process, the structural composition of PtNi/C transforms into PtNi-O/C. PtNi-O/C had the highest mass activity at 70 mV overpotential against the reversible hydrogen electrode compared to commercial Pt/C (0.92 mA/gpt) and PtNi (5.35 mA/gpt), measured against the reversible hydrogen

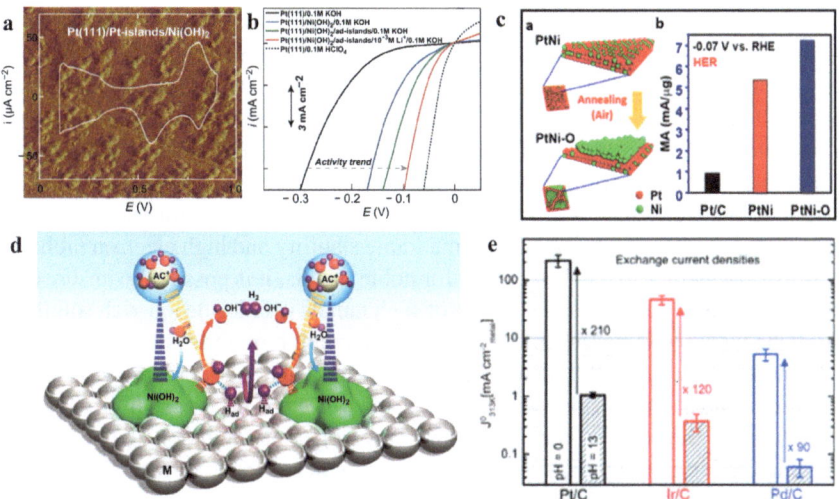

Figure 3.1. (a) STM image (60 nm × 60 nm) and CV trace of the Pt/Ni(OH)$_2$ surface. (b) Comparison of HER activities of Ni(OH)$_2$-modified Pt electrode and control samples in 0.1 M KOH reused with permission from ref [34] copyright 2020 Royal Society of Chemistry. (c) (a) a schematic representation of the annealing in air process that transforms PtNi/C into PtNi-O-C, and (b) evaluation of the HER mass activities at 0 reused with permission from ref [35] copyright 2020 Springer. (d) Schematic representation of water dissociation, formation of M–H$_{ad}$ intermediates, and subsequent recombination of two H$_{ad}$ atoms to form H$_2$ (magenta arrow), as well as OH$^-$ desorption from the Ni(OH)$_2$ domains (red arrows) followed by adsorption of another water molecule on the same site (blue arrows) reused with permission from ref [34] copyright 2020 Royal Society of Chemistry. (e) HOR/HER exchange current densities in PEMFC (empty columns) and 0.1 m NaOH (striped columns) [36]

electrode (RHE). Similar noble metal based electrocatalyst used for hydrogen evolution reaction as shown in Fig. 3.1.

3.2 Carbon Incorporated Noble Metals Based Electrocatalysts

Due to their simple production technique and adaptable surface physicochemical features, carbon-supported metal catalysts (CMCs) are gaining popularity [37]. Modifying the surface of metal particles via contact with carbon can increase their catalytic activity. To distribute metal loading consistently throughout CMC, it is necessary to enhance its surface and interface. Rh nanoparticles loaded onto carbon nanosheets with N/S co-doping were successfully fabricated by Liu et al.) [38]. Rh nanoparticles were uniformly dispersed due to these microscopic mesoporous nanosheets with 437.1 m$_2$/g surface area. In addition, carbon incorporated noble metals based electrocatalysts are shown in Fig. 3.2.

An increase in electron transit to the Rh/carbon interface and an increase in electron ejection from Rh were facilitated by the addition of Sulphur to the light carbon nanosheets, which affects how Rh nanoparticles interacted with the carbon support. This Rh-based catalyst demonstrated remarkably increased catalytic activity in seawater with only a 0.5 weight percent loading, identical to the commercially available 20% Pt/C performance. The catalyst also achieved a constant 15 mA cm2 current density for ten hours. Graphene is an exciting candidate for use as a carrier for carbon-based catalysts due to its remarkable stability and high electron mobility. By using graphene as a catalyst carrier for noble metals, it is possible to ensure high catalytic efficiency and greater stability of such metals in chloride ion-rich solutions, simultaneously reducing the amount of valuable metal required.

Sarno et al. developed a nanostructured catalytic material comprised of nickel, ruthenium, and iridium loaded with graphene [41]. The presence of iridium enhanced the stability of the alloy and ensured a strong hydrogen precipitation activity in both acidic and saline environments. As electrons accumulated on the surface of Ir, activity was further increased by the alloy's composition, which included metals with widely varying work functions. Furthermore, the accumulated negative charges on Ir have protected it from being attacked by chloride anions. The highly conductive graphene network decreased the impedance of charge transmission between the electrode and the nanoparticles trapped on it. After 11,000 cycles in 0.5 M H_2SO_4, the overpotential of 0.06 V and the Tafel slope of 28 mV dec1 remained unchanged [42].

After 250 cycles in actual seawater, there was no measurable change in current density for the samples, which exhibited a Tafel slope of 48 mVdec1. Synthetic electrocatalysts for HER in saltwater were shown to be effective, using synergistic alloying effects and graphene carriers.

3.3 Metal Phosphides Based Electrocatalysts

The HER-stimulating properties of metal phosphides result from the ability of their electronegative P atom to absorb positively charged protons. Metal phosphide can display extraordinary electrical conductivity when the metal-to-phosphorus atom ratio is optimal. Excellent catalytic activity was observed for CoMoP@C at pH $=$ 0–1 and pH $=$ 2–14, with values approaching 20% Pt/C and exceeding 20% Pt/C at high overpotential [43]. The carbon shell's greater ability to absorb protons boosted HER's efficiency. The catalyst in the CoMoP core was shielded from deterioration, clumping, and the corrosive effects of sea water by the carbon coating on the material. Thus, CoMoP@C showed remarkable HER performance in natural seawater. Lv et al. used hydrothermal and phosphorylation methods to produce porous NiCoP electrocatalysts that resembled feathers on foam nickel [44].

The release of the produced Hydrogen was also aided by their conductive substrate and porous architecture, which exposed more active sites and enhanced a specific surface area. Incorporating electrical effects, conductive substrates, and 3D pore patterns enhanced electrochemical stability. In studies conducted in seawater, at a

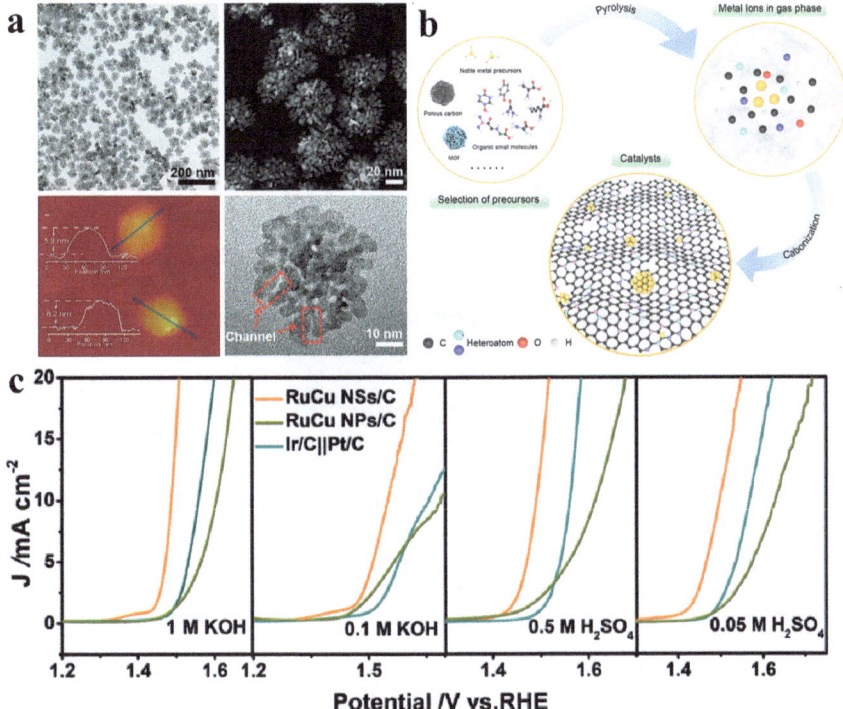

Fig. 3.2 (a) TEM image, HAADF-STEM image, AFM images andcorresponding thickness, high-magnification TEM image. (b) Schematic diagram showing the synthesis principle based on pyrolyzing mixture of precious metal and carbon precursors reused with permission from ref [39] copyright 2022 Wiley. (c) LSV curves of RuCu NSs/C, RuCu NPs/C and Ir/C k Pt/C for overall water splitting in 1 m KOH, 0.1 m KOH, 0.5 m H_2SO_4 and0.05 m H_2SO_4, respectively reused with permission from ref [40] copyright 2019 Wiley

287 mV overpotential and a current density of 10 mA cm^2, NiCoP/NF displayed extraordinarily high levels of stability and activity. SEM micrographs and LSV curves of vanadium and cobalt based materials are presented in Fig. 3.3.

3.4 Metal Oxides and Hydroxides Based Electrocatalysts

In natural saltwater environments, metal oxides can perform similarly to platinum as HER catalysts thanks to their wide diversity of crystal shapes, abundance, and great catalytic activity. Metal oxides as HER catalysts are still in the early stages of industrialization. For Mn-MOF to interact with nickel components, nickel foam was employed as a substrate. Mn-doped nickel oxide/Ni (Mn-NiO/Ni) was produced by pyrolyzing an Mn-MOF/Ni-F precursor in an oxygen-free environment [47].

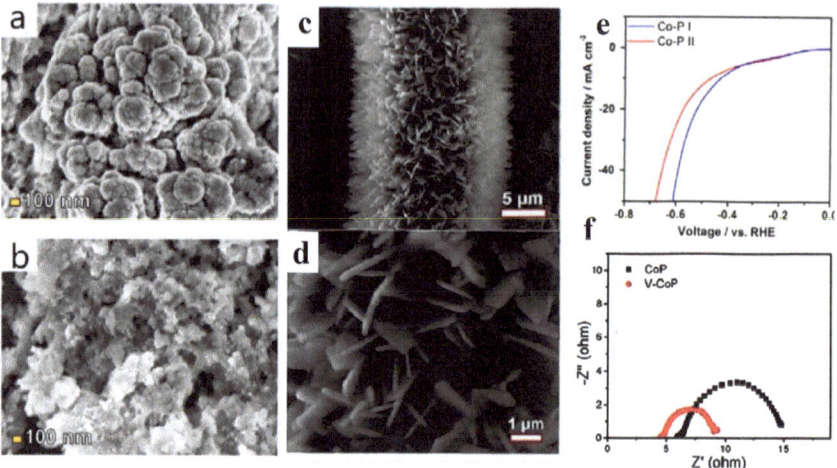

Fig. 3.3 SEM image of (a) as-synthesized Co–P I (b) as-synthesized Co–P II and (e) HER performance of Co–P I and Co–P II in 0.2 mol/L phosphate buffer solution reused with permission from ref [45] copyright 2018 American Chemical Society. c, d and (f) SEM images of CoMOF and V–CoP/CC *vs.* CoP energy Stratgram reused with permission from ref [46] copyright 2020 Frontier in chemistry

Amal et al. found that the degree of Ni oxidation had a major effect on the device's overall performance while developing carbon-based NiO/Ni for HER [47]. When it comes to modifying the electrical structure of the catalyst by doping noble metal atoms in metal oxides, Ru was found to be more cost-effective than other noble metals. Rh is possibly favorable and promising for HER electrocatalysis since its hydrogen binding energy is comparable to that of Pt. Additionally, the metal oxide materials' structural makeup influences their function as electrocatalysts. Amorphous materials might have a lot of exposed surfaces and flaws due to the atom arrangement. The amorphous cobalt oxide Ru-CoOx/NF was synthesized by Wu et al. to contain extremely small amounts of Ru. More active sites for electrocatalytic processes were available because of the amorphous structure.

Additionally, the insertion of Ru components facilitated internal charge transfer, improving performance. These favorable characteristics efficiently drove its electrolysis in seawater media at high current densities. Metal hydroxides, as well as metal oxides, have been used in HER. Electrodeposition on the surface of NF was used by Jiang et al. to construct NiFe-LDH/FeOOH heterostructure nanosheets. By combining NiFe-LDH and FeOOH, the catalyst showed impressive HER performance in a 0.5 M NaCl$^+$ 1.0 M KOH electrolyte [48].

NiFe-LDH/FeOOH had an overpotential of 181.8 mV at a current density of 10 mA cm2 and an overpotential of 286.2 mV at a current density of 100 mA cm2. At a current density of 100 mA cm2, NiFe-LDH/FeOOH displayed significant catalytic activity and high stability levels over 105 h. Additionally, it was shown that cation doping could alter the total three-dimensional energy of electrocatalysts and change the electrical composition of intermediates and the surface adsorption energy of those intermediates [48].

3.5 Transition Metal Carbides (TMCs) Based Electrocatalysts

The discovery of non-noble metal-based electrocatalysts has increased interest in transition metal carbides (TMCs). Mo_2C and tungsten carbide (WC) are two types of materials that are examples of substances that have an improved catalytic activity for HER. Along with the materials' high electrical conductivity, the d-band electronic density state (like Pt) and hydrogen adsorption qualities are likely the key sources of the observed considerable HER activity. Both of these characteristics are similar to Pt. Levy and Boudart discovered that tungsten carbide had platinum-like catalytic activity in 1973 and shared d-band electronic density states with Pt species [49]. Additionally, Chen et al. used DFT calculations to examine several transition metal carbides' physical, chemical, and electronic structural characteristics [50]. Carbon atoms inserted into lattice interstices were found to provide d-band electronic density states comparable to the Pt reference.

Experiments couldn't verify the theory's prediction until 2012. Hu et al. demonstrated the potent HER catalytic activity of commercially available microparticles of molybdenum carbide (com-Mo2C) under both acidic and basic environments [51]. However, a significant overpotential is needed for a cathodic current of 10 mA cm^2 (190–230 mV). Researchers studied various strategies for the Mo2C catalyst optimization by Nano engineering the materials to expose additional active areas in response to the groundbreaking investigations. By easily carburizing anilinium molybdate in hydrogen, Wang et al. were able to successfully create Mo_2C nanorods with a porous structure, as expressed in. Field emission scanning electron microscopy (FE-SEM) and Transmission electron microscopy (TEM) images were used to capture the nanorod morphologies with a porous structure and smooth surface, respectively [52].

For the Mo_2C nanorod catalyst, improved HER electrocatalytic activity is demonstrated. The linear scan voltammetry (LSV) consequences in 0.5 M H_2SO_4, proving that the Mo_2C nanorods are more active than their commercial counterparts. The outcomes illustrate that the activity has not changed after two thousand cycles, pointing to a long cycling life. Due to their strong conductivity and well-defined and porous structure, Mo2C nanorods are a potential choice for HER that can be used in both acidic and alkaline situations. Additionally, the Mo_2C nanorod is evaluated in alkaline conditions, with 1 M KOH showing improved performance than commercial Mo_2C. Ni nanoparticle addition could enhance the catalytic activity even more.

Coatings of molybdenum carbide can be added to carbon-based materials further to enhance the HER's performance as a hybrid nano-electrocatalyst. Mo_2C nanoparticles were produced by Liu et al. by hydrothermally synthesizing carbides on the GNR template and then calcining them at high temperatures (GNRs), similar mechanism is shown in Fig. 4.

In all neutral, basic, and acidic environments, the Mo_2C-GNR hybrid demonstrates exceptional electrocatalytic activity and durability. Because the linked GNR network topology provides various conductive channels for rapid electron transport, an abundance of exposed active sites, and large accessible surface areas, the catalytic activity may increase across all the different media types, including acidic, basic, and neutral. Because of this, the promising concept of using GNRs as models for in-situ carbide manufacturing arises.

3.6   Transition Metal Phosphides Based Electrocatalysts

Research on transition metal phosphides (TMP) is growing at a quick pace as scientists strive to develop electrocatalysts that have good catalytic potential and are stable in both acidic and alkaline conditions (pH universal). Due to its exceptional conductivity and distinctive electronic structure, it has been suggested that the P atom contributes significantly to the formation of TMP. Ni_2P catalysts were first identified as one of the finest real-world HER catalysts in 2005. Liu and Rodriguez used density functional theory to explore a variety of electrocatalysts (DFT). The findings demonstrate HER activity in the [NiFe] hydrogenase > [Ni (PNP)$_2$] trend. Ni2P (001) > 2 + > [Ni (PS3*) (CO)] 1 − > Pt > Ni. Due to the hydrogen's strong bonding to the hollow metal (Pt, Ni) sites created during the HER process, Ni2P exhibits greater activity than bulk Pt and Ni. Due to the tight connection, the hydrogen species would require more energy to desorb from the metal surface. Bonding to intermediates and products is moderate because of P atoms, which decreases the number of potentially active Ni sites (also known as the "ensemble effect").

With an exchange density of 3.3×105 mA cm^2 and a Tafel slope of 46 m Vdec1, the HER activity of Ni_2P nanoparticles that were placed onto a Ti foil substrate was exceptional. This was the first proof of the catalytic synergy obtained through direct experimental testing. The Ni_2P/Ti electrode's stability was insufficient, particularly in an alkaline electrolyte. To further their investigation, Hu et al. developed $NiCo_2Px$, a phosphide electrocatalyst with a bimetallic structure. This catalyst is robust and stable over extended periods in a wide range of electrolytes, and it functions well as a pH-universal catalyst for HER. The wet chemical-hydrothermal method and in-place phosphorization reaction on commercial carbon felt were used to construct self-supported $NiCo_2Px$ nanowire arrays (CF) [55].

Figure 3.5. demonstrates $NiCo_2Px$'s superior HER performance in neutral, basic, and acidic conditions. In basic electrolytes, the overpotential of $NiCo_2Px$ is the lowest, coming in at 58 mV at 10 mA cm^2 when compared to NiPx (180 mV), CoPx (94 mV), and commercial Pt. This is because $NiCo_2Px$ has a lower oxidation potential

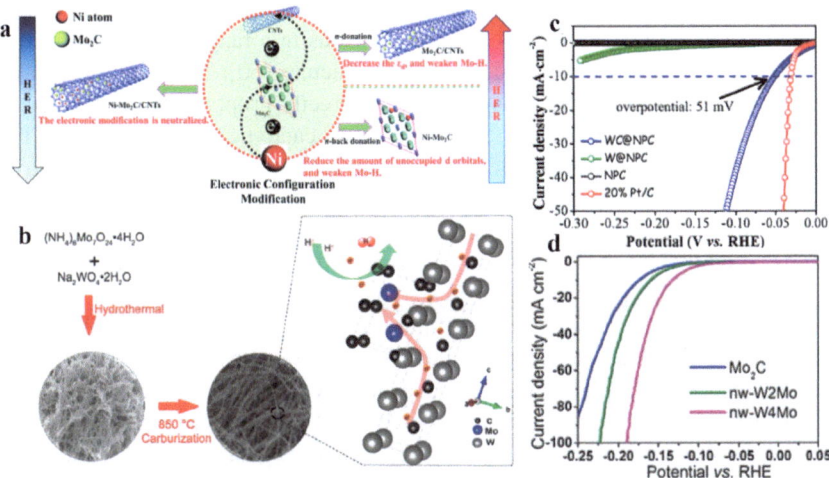

Fig. 3.4 (a) Schematic diagram showing the effects of electron-donating and electron-withdrawing materials together with the combined effect. (b) Schematic formation of molybdenum carbide–tungsten carbide composite nanowires reused with permission from ref [53] copyright 2015 Wiley. (c) LSV curves of WC@NPC and several materials reused with permission from ref [54] copyright 2017 American Chemical Society. (d) Linear sweep voltammograms of various samples after CV activation reused with permission from ref [53] copyright 2015 Wiley

(70 mV). As shown by the long-term HER activity test results, the morphology of $NiCo_2Px$ reveals that the catalyst structure has been maintained in an excellent state even after 5000 cycles under a diverse set of conditions.

For $NiCo_2PX$ in an alkaline electrolyte, a synergistic effect is suggested. In this molecule, the particularly significant interactions are the ones between the hanging the H atom and P atom (P), as well as the contacts between the center of under-coordinated metal (M + , M = Co, Ni) and the O atom. The interaction of these two forces finally causes the water molecule to break up into its component pieces, which is what we mean when we talk about dissolving (an OH molecule and an H atom). After that, the H+ atom is moved to a nearby metal site that is now empty, where it will continue to exist as adsorbed hydrogen after the transfer. After combining to generate H_2, the adsorbed H atoms are subsequently transferred to an available metal site [58].

3.7 Transition Metal Chalcogenides Based Electrocatalysts

Using DFT calculations, Nrskov et al. (2005) showed that the free energy of hydrogen atoms bonding to the edge of MoS_2 is comparable to that of Pt, verified by the researchers [59]. Based on these findings, it would suggest that MoS_2 can serve as an electrocatalyst that is helpful for HER. Chorkendorff and colleagues created Au (111)-supported triangular MoS_2 single crystals in a range of sizes to localize the action's location within the MoS_2 crystal structure [60]. They showed that the quantity of edge sites on the MoS_2 catalyst is directly proportional to the electro-catalytic HER activity. MoS_2 has extremely active catalytic sites at its edges. This understanding has led to the development of numerous methods, such as nanostructure tuning and morphology tuning, to expose the active areas and increase HER activity. An approach to design flaws in MoS_2 ultrathin nanosheets, for instance, was disclosed by Xie et al. and was demonstrated to significantly enhance MoS_2's electrocatalytic HER performance [61].

The rich-defect structure generated more active edge sites responsible for the enhanced activity. These active edge sites were generated by partially severing the catalytically inert basal plane. Jin et al. independently achieved the effective synthesis of CoS_2 in the form of adaptable films, nanowires (NW), and microwires (MW). They assessed the structures, stabilities, and activities of the organisms and concluded that the observed increase in activity and stability is due to better morphology. [62]. This was the result of their research after they analyzed their structures, activities, and stabilities.

In terms of HER catalytic efficiency and stability, A relatively large electrode surface area and rapid release of generated gas bubbles from the electrode surface set CoS_2 NWs apart from the other two morphologies. Significant work has been put towards tailoring metal chalcogenides' electronic conductivity and their active sites to maximize the HER electrocatalytic activity. The use of heteroatom doping to increase HER activity is efficient. According to the findings of Xie and colleagues, the electrical structure of MoS_2 ultrathin nanosheets may be successfully modified by incorporating oxygen atoms and adjustable engineering disorder [63]. Consequently, HER activity and conductivity both increase. The MoS_2 ultrathin nanosheets produced at a high temperature (220 °C) displayed numerous defects, which displays a sequence of XRD patterns for the catalysts created at different temperatures.

When the synthesis temperature is lowered to 200 °C or below, two new peaks appear in the low-diffraction angle area, revealing an unusual lamellar structure with an enhanced interlayer spacing of 9.5 compared to 6.15 in pure $2H-MoS_2$. FFT patterns and HRTEM images of the MoS_2 structures at varying temperatures show that the disorder may be engineered in a controlled manner. Several transition metal chalcogenides based electrocatalysts for hydrogen evolution reactions are depicted in Fig. 3.6.

The degree of chaos is temperature dependent. Electrochemical research on ultrathin MoS_2 nanosheets with varying degrees of disorder. These experiments were carried out with oxygen incorporation (b).

The S180 catalyst has the lowest potential (120 mV), which indicates that its HER activity is superior to the other two. A schematic illustration of the disordered structure that can be seen in ultrathin nanosheets of MoS_2 with oxygen incorporation can be seen in demonstrates that electrons move quite quickly between the quasi-periodically aligned nanodomains (e). Because of the chaos they bring into the environment, these areas contribute to enrichment in a positive way (purple shading). The

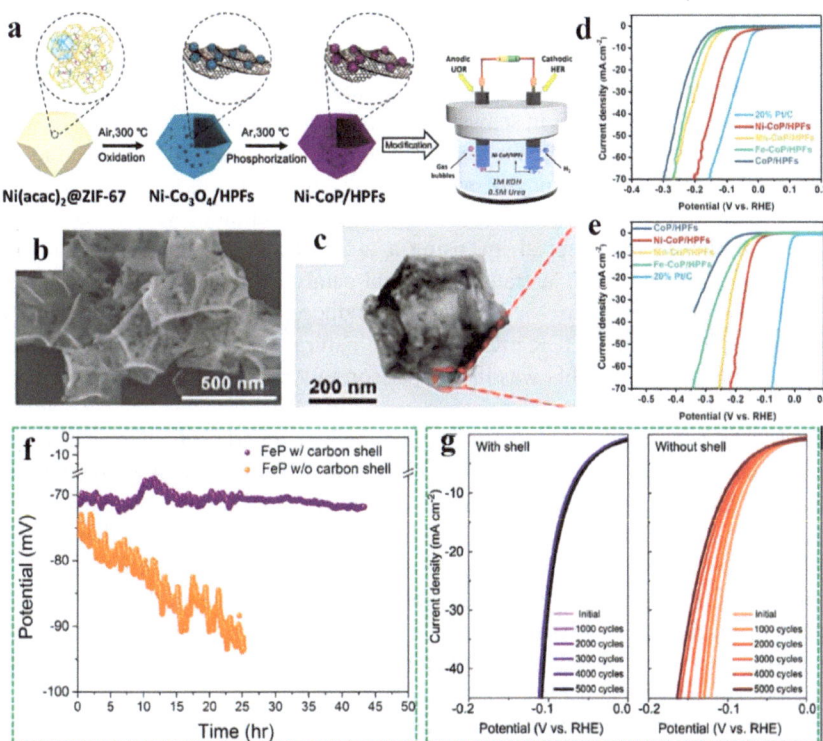

Fig. 3.5 (a) Synthesis scheme, (b) SEM and (c) TEM images of Ni-CoP/HPFs. (d) and (e) LSV curves in 0.5 M H$_2$SO$_4$ and 1 M KOH reused with permission from ref [56] copyright 2019 Elsevier. (f) Chronopotentiometry results and (g) LSV curves of FeP nanoparticles with (left) and without (right) a carbon shell reused with permission from ref [57] copyright 2017 American Chemical Society

disordered structure of HER, which offers many unsaturated sulfur atoms as active sites and promotes the quasiperiodic structuring of nanodomains required for quick electron transit, is responsible for the material's remarkable electrocatalytic activity. This is the case because it allows rapid electron transit. This schematic displays the unorganized structure of ultrathin nanosheets of MoS$_2$ that have oxygen incorporated into their structure. Because of the chaotic structure, rapid electron transit across the quasi-periodically aligned nanodomains is illustrated by purple smears on blue lines, indicating an enrichment influence of active sites.

3.8 Metal Nitrides Based Electrocatalysts

Due to their superior corrosion resistance and high conductivity, metal nitrides have shown considerable promise in seawater electrolysis [66]. Techniques such as interface engineering, hetero-element doping, vacancy engineering, and alloying can improve the catalytic efficiency of metal nitrides and compensate for the inherent flaws present in metal nitrides. The atomic scale production of Mo_5N_6 nanosheets was accomplished by Jin et al. by using a two-dimensional lateral growth technique in conjunction with a transition metal-catalyzed phase transition strategy [67]. Two-dimensional Mo_5N_6 nanosheets produced through rich metal-nitrogen bonding depicted higher HER presentation in natural seawater and stability under high currents for 100 h. This was in comparison to typical nitrogen-deficient metal nitrides.

In comparison to other metal nitrides and the benchmark Pt/C, the performance was noticeably superior. Because of its electrical structure, which is analogous to platinum, Mo_5N_6 possesses a high activity level. Its stability can be related to the fact that the Mo atoms included within it have a high valence, making them far less sensitive to active-site poisoning than other ions in saltwater, which is why it is so stable. In addition, the ratio of nitrogen atoms to metal atoms in the matrix of the metal nitride can be altered to influence the electrical structure of the metal nitride.

Nitrogen enrichment and incomplete nitration processes were used to manage the nitrogen content of metal nitrides. Nitrogen enrichment is a method developed to increase the abundance of nitrogen atoms in the lattice of metal nitrides; however, it typically necessitates high temperatures and pressures for optimal results. The creation of the metal/metal nitridation interface, which is normally more electrically conductive and electro-catalytically active, can be helped along by incomplete nitridation of the metals involved. Unsaturated nitridation was used by Jin et al. to create nickel surface nitride with a carbon shell, which they referred to as Ni–SN@C.

Ni–SN@C was unique because, unlike other metal/metal nitrides and transition metal nitrides, its heterostructures lacked the bulk nickel nitride phase with surface unsaturated NiN bonds. In contrast, Ni–SN@C was mostly composed of nickel metal and included a novel unsaturated NiN bond on its surface. At 10 mA cm2, the overpotential of the Ni–SN@C catalyst in alkaline seawater was just 23 mV [67]. HER performance regarding metal nitride based electrocatalysts are depicted in Fig. 3.7.

This technique resulted in catalysts with the appropriate metal nitride characteristics and promoted the rearrangement of charges on the catalyst's surface, resulting in high activity for HER in seawater electrolysis and excellent corrosion resistance. Metal oxides, hydroxides, nitrides, phosphides, and other compounds have successfully catalyzed the creation of hydrogen in seawater electrolytes. All of the resulting heterogeneous compounds have similar catalytic potential. Yu et al. created a sandwich-like array of nickel phosphide (NixP) microplates by manipulating the strength of electrical connections between two chemicals, fine-tuning the charge distribution within the hybrid electrocatalyst, and optimizing the energy of hydrogen

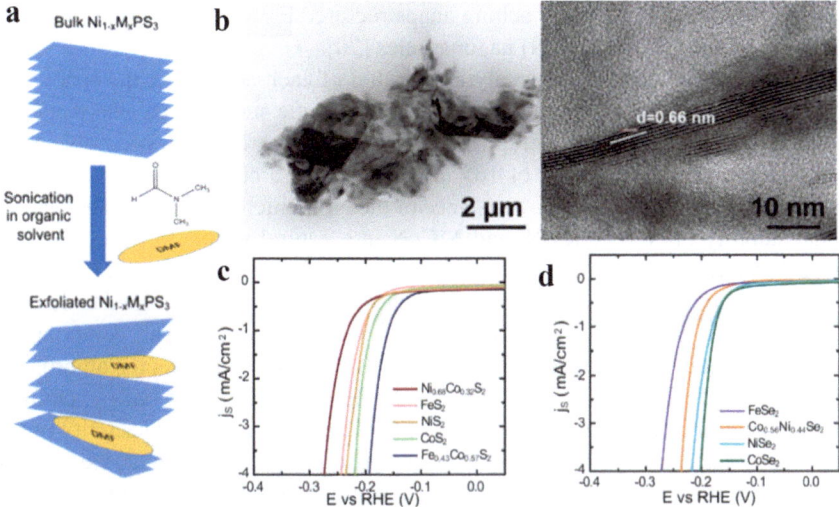

Fig. 3.6 (a) Schematic illustration of $Ni_{1-x}M_xPS_3$ (where M is a doping metal) nanosheet preparation by exfoliation of bulk crystals in organic solvent (N,N-dimethylformamide (DMF)), (b) TEM micrographs of $Ni_{1-x}M_xPS_3$ reused with permission from ref [64] copyright 2018 Elsevier. (c) LVS curves of transition metal disulphides in which surface-area-normalized current density is plotted against the potential in 0.5 M H_2SO_4. (d) LVS curves of transition metal diselenides in which surface-area-normalized current density is plotted against the potential in 0.5 M H_2SO_4 reused with permission from ref [65] copyright 2013 Royal Society of Chemistry

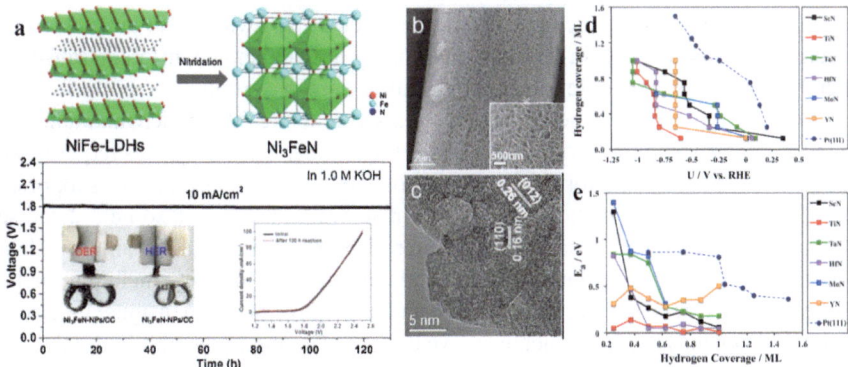

Fig. 3.7 (a) 3D structure of NiFe-LDHs and Ni_3FeN. and Time and Voltage plot for HER and OER, (b-c) SEM image, HRTEM of NiFe-LDHs nanosheets reused with permission from ref [68] copyright 2017 Elsevier. (d) Hydrogen coverage on the surface of the transition-metal nitrides as a function of the applied potential (U/V vs RHE). (e) Activation energy (Ea/eV) for the Tafel reaction in HER on several transition-metal nitrides as a function of the hydrogen coverage reused with permission from ref [69] copyright 2017 American Chemical Society

adsorption. They did this by attaching nanostructured HER catalysts to the surfaces of nickel cobalt nitride (NiCoN) nanoparticles [70].

In addition to boosting the catalytic potential of each active site, the sandwich-shaped catalyst's (NiCoN| NixP| NiCoN) large specific surface area also made for good conductivity, allowing for rapid and efficient charge transfer. It was only necessary to apply an overpotential of 165 mV to generate a current density of 10 mA cm2 thanks to the high chloride resistance of the NixP microplate array. This exceptional stability of NixP, NiCoN, and NiCoN in a natural saltwater electrolyte was made possible since NiCoN| NixP| NiCoN could exist in this electrolyte. Treating the nickel precursor in a nitrogen environment was all Huang et al. needed to develop a hybrid electrocatalyst that significantly increased HER activity. This compound is composed of the hydrogen acceptor Ni_5P_4 and the hydroxyl acceptor amorphous nickel-hydrogen (oxygen) oxide [Ni2 + O (OH) 2]. Ni_5P_4 and Ni2 + O(OH)$_2$ hybridization increased the specific surface area and electrocatalytic activity of the composite electrocatalyst by enhancing water adsorption, improving the free energy of hydrogen adsorption, and activating the catalytic pathway over a wide pH range [5]. The constructed hybrid catalysts showed overpotentials of 108 mV dec.1 in actual seawater and 144 mV at 10 mA cm^2. The increased efficiency was linked back to the synergistic effect of the phosphide, the increased specific surface area, and the interaction between the hydroxide and the electrical component.

Hydrogen evolution reaction (HER) generally encounters fewer competitive responses than OER. However, contaminants in saltwater harm the HER reaction mechanism, leading to undesirable outcomes such as insoluble precipitation. The most notable HER catalysts at the current time are still Pt and other noble metals, although Pt's expensive price prevents it from being used widely. So far, numerous studies on the use of HER in seawater electrolysis have examined different transition metal compounds and noble metal alloys with promising results.

Pt can be alloyed with other transition metals to generate PtM, which maintains the catalytic activity of pure Pt while decreasing the number of noble metals required for the reaction. M can stand for Cr, Fe, Co, Ni, or Mo. Not only can catalysts supported by carbon make noble metals more stable in seawater, but they also help disperse the metals more evenly throughout the water. High HER presentation and excellent resistance to saltwater pollutants are features of other compounds, including metal oxides (hydroxides), nitrides, phosphides, and innovative two-dimensional materials termed MXenes. The Nobel metals we've discussed so far are in addition to these.

References

1. Trasatti S (1972) Work function, electronegativity, and electrochemical behaviour of metals: III. Electrolytic hydrogen evolution in acid solutions. J Electroanal Chem Interfacial Electrochem 39(1): 163–184
2. Wang S, Lu A, Zhong C-J (2021) Hydrogen production from water electrolysis: role of catalysts. Nano Convergence 8(1):1–23

3. Kakaei K, Esrafili MD, Ehsani A (2019) Alcohol oxidation and hydrogen evolution. Interface Science and Technology. Elsevier, pp 253–301

4. Jiang S et al (2022) Recent advances in seawater electrolysis. Catalysts 12(2):123

5. Huang Y et al (2019) Nitrogen treatment generates tunable nanohybridization of Ni5P4 nanosheets with nickel hydr (oxy) oxides for efficient hydrogen production in alkaline, seawater and acidic media. Appl Catal B 251:181–194

6. Sarno M, Ponticorvo E, Scarpa DJEC (2020) Active and stable graphene supporting trimetallic alloy-based electrocatalyst for hydrogen evolution by seawater splitting. Electrochem Commun 111:106647

7. Shang L et al (2020) Underwater superaerophobic Ni nanoparticle-decorated nickel–molybdenum nitride nanowire arrays for hydrogen evolution in neutral media. Nano Energy 78:105375

8. Loh A et al (2020) Development of Ni–Fe based ternary metal hydroxides as highly efficient oxygen evolution catalysts in AEM water electrolysis for hydrogen production. Int J Hydrogen Energy 45(46):24232–24247

9. Zhang D et al (2020) Design and in-situ synthesis of unique catalyst via embedding graphene oxide shell membrane in NiS2 for efficient hydrogen evolution. Appl Surf Sci 510:145483

10. Kim C et al Cobalt–iron–phosphate hydrogen evolution reaction electrocatalyst for solar-driven alkaline seawater electrolyzer. Nanomaterials 11(11):2989

11. Yang T et al (2021) Triggering the intrinsic catalytic activity of Ni-doped molybdenum oxides via phase engineering for hydrogen evolution and application in Mg/seawater batteries. ACS Sustain Chem Eng 9(38):13106–13113

12. Liu D et al (2021) Multi-phase heterostructure of CoNiP/CoxP for enhanced hydrogen evolution under alkaline and seawater conditions by promoting H2O dissociation. Small 17(17):2007557

13. Zhang B et al (2021) Enhanced interface interaction in Cu2S@ Ni core-shell nanorod arrays as hydrogen evolution reaction electrode for alkaline seawater electrolysis. J Power Sources 506:230235

14. Liu G et al (2021) Porous CoP/Co2P heterostructure for efficient hydrogen evolution and application in magnesium/seawater battery. J Power Sources 486:229351

15. Lyu C et al (2022) Interfacial electronic structure modulation of CoP nanowires with FeP nanosheets for enhanced hydrogen evolution under alkaline water/seawater electrolytes. Appl Catal B 317:121799

16. Wu L et al (2022) Efficient alkaline water/seawater hydrogen evolution by a nanorod-nanoparticle-structured Ni-MoN catalyst with fast water-dissociation kinetics. Adv Mater 34(21):2201774

17. Bu X et al (2022) NiMo@ C3N5 heterostructures with multiple electronic transmission channels for highly efficient hydrogen evolution from alkaline electrolytes and seawater. Chem Eng J 438:135379

18. Sun H et al Nickel–cobalt hydrogen phosphate on nickel nitride supported on nickel foam for alkaline seawater electrolysis. ACS Appl Mater Interfaces 14(19):22061–22070

19. Meng G et al (2022) Co—W bimetallic carbide nanocatalysts: computational exploration, confined disassembly-assembly synthesis and alkaline/seawater hydrogen. Evolution 18(48):2204443

20. Xu Y et al (2022) Mg/seawater batteries driven self-powered direct seawater electrolysis systems for hydrogen production. Nano Energy 98:107295

21. Vijayapradeep S et al. (2023). Novel Pt-carbon core–shell decorated hierarchical CoMo2S4 as efficient electrocatalysts for alkaline/seawater hydrogen evolution reaction. Chem Eng J 473:145348

22. Chen S et al (2023) Multi-metal electrocatalyst with crystalline/amorphous structure for enhanced alkaline water/seawater hydrogen evolution. J Colloid Interface Sci 650:807–815

23. Nguyen N-A et al (2023) High electrocatalytic activity of Rh-WO3 electrocatalyst for hydrogen evolution reaction under the acidic, alkaline, and alkaline-seawater electrolytes. Int J Hydrogen Energy 48(84):32686–32698

24. Dang Y et al (2023) Enhanced alkaline/seawater hydrogen evolution reaction performance of NiSe2 by ruthenium and tungsten bimetal doping. Int J Hydrogen Energy 48(45):17035–17044

25. Barua S et al (2024). 3D Nickel–Manganese bimetallic electrocatalysts for an enhanced hydrogen evolution reaction performance in simulated seawater/alkaline natural seawater. Int J Hydrogen Energy 79:1490–1500

26. Geng, S., et al., Synergistic dp hybridized Co/W5N4 heterostructure catalyst for industrial alkaline water/seawater hydrogen evolution. Appl Catal B 343:123486

27. Grigoriev S, Millet P, Fateev V (2008) Evaluation of carbon-supported Pt and Pd nanoparticles for the hydrogen evolution reaction in PEM water electrolysers. J Power Sources 177(2):281–285

28. Renjith A, Roy A, Lakshminarayanan V (2014) In situ fabrication of electrochemically grown mesoporous metallic thin films by anodic dissolution in deep eutectic solvents. J Colloid Interface Sci 426:270–279

29. Fang Y-H, Liu Z-P (2010) Mechanism and tafel lines of electro-oxidation of water to oxygen on RuO2 (110). J Am Chem Soc 132(51):18214–18222

30. Xie L et al (2018) Superior alkaline hydrogen evolution electrocatalysis enabled by an ultra-fine PtNi nanoparticle-decorated Ni nanoarray with ultralow Pt loading. Inorganic Chemistry Frontiers 5(6):1365–1369

31. Zheng J (2017) Seawater splitting for high-efficiency hydrogen evolution by alloyed PtNix electrocatalysts. Appl Surf Sci 413:360–365

32. Subbaraman R et al (2011) Enhancing hydrogen evolution activity in water splitting by tailoring Li+-Ni (OH) 2-Pt interfaces. Science 334(6060):1256–1260

33. Zhao Z et al (2018) Surface-engineered PtNi-O nanostructure with record-high performance for electrocatalytic hydrogen evolution reaction. J Am Chem Soc 140(29):9046–9050

34. Chen M et al (2020) The coupling of experiments with density functional theory in the studies of the electrochemical hydrogen evolution reaction. Journal of Materials Chemistry A 8(18):8783–8812

35. Wu Y, Yao J, Gao J (2020) Interface chemistry of platinum-based materials for electrocatalytic hydrogen evolution in alkaline conditions. Methods Electrocatal: Adv Mater Allied Appl 453–473.

36. Tang T et al (2022) Synergistic electrocatalysts for alkaline hydrogen oxidation and evolution reactions. Adv Func Mater 32(2):2107479

37. Yang S-C et al (2018) Synergy between ceria oxygen vacancies and Cu nanoparticles facilitates the catalytic conversion of CO2 to CO under mild conditions. ACS Catal 8(12):12056–12066

38. Liu Y et al (2019) Surface engineering of Rh catalysts with N/S-codoped carbon nanosheets toward high-Performance hydrogen evolution from seawater. ACS Sustain Chem & Eng 7(23):18835–18843

39. Liu Y et al (2022) Recent advances in carbon-supported noble-metal electrocatalysts for hydrogen evolution reaction: syntheses, structures, and properties. Adv Energy Mater 12(28):2200928

40. Yao Q et al (2019) Channel-rich RuCu nanosheets for pH-universal overall water splitting electrocatalysis. Angew Chem 131(39):14121–14126

41. Sarno M et al (2012) Evaluating the effects of operating conditions on the quantity, quality and catalyzed growth mechanisms of CNTs. J Mol Catal A: Chem 357:26–38

42. Sarno M, Ponticorvo E, Scarpa D (2020) Active and stable graphene supporting trimetallic alloy-based electrocatalyst for hydrogen evolution by seawater splitting. Electrochem Commun 111:10664

43. Ma Y-Y et al (2017) Highly efficient hydrogen evolution from seawater by a low-cost and stable CoMoP@ C electrocatalyst superior to Pt/C. Energy Environ Sci 10(3):788–798

44. Lv Q et al (2019) Featherlike NiCoP holey nanoarrys for efficient and stable seawater splitting. ACS Applied Energy Materials 2(5):3910–3917

45. Sumboja A et al (2018) One-step facile synthesis of cobalt phosphides for hydrogen evolution reaction catalysts in acidic and alkaline medium. ACS Appl Mater & Interfaces 10(18):15673–15680

46. Hua W et al (2020) V-Doped CoP nanosheet arrays as highly efficient electrocatalysts for hydrogen evolution reaction in both acidic and alkaline solutions. Front Chem 8:608133
47. Lovell EC et al (2020) From passivation to activation–tunable nickel/nickel oxide for hydrogen evolution electrocatalysis. Chem Commun 56(11):1709–1712
48. Jiang K et al (2021) NiFe layered double hydroxide/FeOOH heterostructure nanosheets as an efficient and durable bifunctional electrocatalyst for overall seawater splitting. Inorg Chem 60(22):17371–17378
49. Levy R, Boudart M (1973) Platinum-like behavior of tungsten carbide in surface catalysis. science 181(4099):547–549
50. Kitchin JR et al (2005) Trends in the chemical properties of early transition metal carbide surfaces: a density functional study. Catal Today 105(1):66–73
51. Vrubel H, Hu X (2012) Molybdenum boride and carbide catalyze hydrogen evolution in both acidic and basic solutions. Angew Chem Int Ed Engl 51(ARTICLE):12703–12706
52. Xiao P et al (2014) Investigation of molybdenum carbide nano-rod as an efficient and durable electrocatalyst for hydrogen evolution in acidic and alkaline media. Appl Catal B 154:232–237
53. Xiao P et al (2015) Novel molybdenum carbide–tungsten carbide composite nanowires and their electrochemical activation for efficient and stable hydrogen evolution. Adv Func Mater 25(10):1520–1526
54. Xu Y-T et al (2017) Cage-confinement pyrolysis route to ultrasmall tungsten carbide nanoparticles for efficient electrocatalytic hydrogen evolution. J Am Chem Soc 139(15):5285–5288
55. Popczun EJ et al (2013) Nanostructured nickel phosphide as an electrocatalyst for the hydrogen evolution reaction. J Am Chem Soc 135(25):9267–9270
56. Pan Y et al (2019) Electronic structure and d-band center control engineering over M-doped CoP (M= Ni, Mn, Fe) hollow polyhedron frames for boosting hydrogen production. Nano Energy 56:411–419
57. Chung DY et al (2017) Large-scale synthesis of carbon-shell-coated FeP nanoparticles for robust hydrogen evolution reaction electrocatalyst. J Am Chem Soc 139(19):6669–6674
58. Zhang R et al (2017) Ternary $NiCo_2Px$ nanowires as pH-universal electrocatalysts for highly efficient hydrogen evolution reaction. Adv Mater 29(9):1605502
59. Hinnemann B et al (2005) Biomimetic hydrogen evolution: MoS_2 nanoparticles as catalyst for hydrogen evolution. J Am Chem Soc 127(15):5308–5309
60. Jaramillo TF et al (2007) Identification of active edge sites for electrochemical H_2 evolution from MoS_2 nanocatalysts. science 317(5834):100–102
61. Xie J et al (2013) Defect-rich MoS_2 ultrathin nanosheets with additional active edge sites for enhanced electrocatalytic hydrogen evolution. Adv Mater 25(40):5807–5813
62. Faber MS et al (2014) High-performance electrocatalysis using metallic cobalt pyrite (CoS2) micro-and nanostructures. J Am Chem Soc 136(28):10053–10061
63. Xie J et al (2013) Controllable disorder engineering in oxygen-incorporated MoS_2 ultrathin nanosheets for efficient hydrogen evolution. J Am Chem Soc 135(47):17881–17888
64. Rakov D et al (2018) Insight into Mn and Ni doping of Ni1-xMnxPS3 and Mn1-xNixPS3 nanosheets on electrocatalytic hydrogen and oxygen evolution activity. J Alloy Compd 769:532–538
65. Kong D et al (2013) First-row transition metal dichalcogenide catalysts for hydrogen evolution reaction. Energy Environ Sci 6(12):3553–3558
66. Pi Y et al (2017) Trimetallic oxyhydroxide coralloids for efficient oxygen evolution electrocatalysis. Angew Chem Int Ed 56(16):4502–4506
67. Jin H et al (2018) Single-crystal nitrogen-rich two-dimensional Mo5N6 nanosheets for efficient and stable seawater splitting. ACS Nano 12(12):12761–12769
68. Chen Q et al (2017) Bifunctional iron–nickel nitride nanoparticles as flexible and robust electrode for overall water splitting. Electrochim Acta 247:666–673
69. Abghoui Y, Skúlason E (2017) Hydrogen evolution reaction catalyzed by transition-metal nitrides. J Phys Chem C 121(43):24036–24045
70. Yu L et al (2020) Hydrogen generation from seawater electrolysis over a sandwich-like NiCoN| Ni x P| NiCoN microsheet array catalyst. ACS Energy Lett 5(8):2681–2689

Chapter 4
Electrocatalyst Design for Oxygen Evolution Reaction

Abstract As discussed earlier, chlorine evolution reaction, makes seawater electrolysis more intriguing than that of freshwater electrolysis due to competition with the oxygen evolution reaction (OER) at the anode. Therefore, significant effort has been devoted to engineering the selective OER electrocatalyst to enhance performance and suppress the chlorine evolution reaction. Based on this, in this chapter, an overview of the latest developments in OER electrocatalysts for seawater splitting was outlined with a highlighting on their significant selectivity performance.

The oxygen evolution reaction (OER) is a limiting reaction in chemical processes that produce molecular oxygen, such as the oxidation of water during oxygenic photosynthesis, the electrolysis of water into hydrogen and oxygen, and the electrocatalytic evolution of oxygen from oxoacids and oxides [1]. The growth of various renewable energy technologies, including solar fuel production and metal-air batteries, depends on the creation of better catalysts for the OER. However, rare and expensive precious metal oxides like iridium and ruthenium exhibit the best OER activity; these factors constrain their scalability. OER catalyst development has focused on first-row metals since they jeopardize noncritical elements. Manganese oxides have been the go-to material to increase gas evolution efficiency because of their stability under oxygen evolution circumstances [2]. In energy storage and conversion systems like regenerative fuel cells, rechargeable metal-air batteries, and electrolyzers, the oxygen evolution reaction (OER) is a fundamental component. However, OER catalysts are forced to operate with a significant intrinsic overpotential and slow reaction kinetics due to the adsorption energy scaling equations between the reaction intermediates [3].

Developing high-activity, stable, and sophisticated electrocatalysts based on non-noble metal materials remains a formidable task. The creation and comprehension of quantitative structure–activity relationships, which relate the catalytic activities to structural and electronic descriptors, is essential to rationalizing innovative and high-efficiency catalysts [4]. To provide a thorough knowledge of the causes of the OER's electrocatalytic activity and further contribute to the development of the electrocatalysis theory, this paper covers the benchmark descriptors for OER electrolysis in great detail. A summary of the most recent research areas for new OER paradigm proposals and critical techniques to get around the scaling connection is provided. The discussion of challenges, possibilities, and views aims to advance the creation of more effective catalysts for improving OER performance and offer some light on the rational design approaches [5].

The numerous chemical steps in OER have traditionally been described using the adsorbate evolution mechanism (AEM). AEM produces oxygen molecules from water in both alkaline and acidic circumstances. The reaction typically involves four coordinated electron and proton transfers, with the metal centers serving as the active site (M).

The red dotted line represents the alkaline OER reaction pathway, which includes steps 4a through 4d:

$$OH^- + M \rightarrow M - OH + e^- \tag{4a}$$

$$M - OH + OH^- \rightarrow M - O + H_2O + e^- \tag{4b}$$

$$M - O + OH^- \rightarrow M - OOH + \frac{e^-}{2M} - O \rightarrow 2M + O_2 + 2e^- \tag{4c}$$

$$M - OOH + OH^- \rightarrow O_2 + H_2) + e^- + M \tag{4d}$$

Many compounds as electro-catalysts, such as d-block metal oxides, metal borides phosphides, and nitrides, are investigated as selective OER catalysts for the electrolysis of seawater (Table 4.1).

4.1 Metal Hydroxides and Oxides Based Electrocatalysts

Manganese-molybdenum oxide electrodes made by Fujimura et al. via anodic deposition on metal hydroxides and oxides are commonly used as OER catalysts due to their greater catalytic properties. In 1998, Izumiya et al. studied the effects of his OER efficiencies of anodized manganese and manganese oxide on tungsten surfaces [25]. This study was devoted to checking pH response and amount of tungsten in oxide anodic deposition in OER. IrO_2 -coated substrate had an OER effectiveness of about 100% at 12 pH in 0.5 M NaCl [26]. Abdel Ghani et al. have determined the OER efficiencies very close to 100% in 0.5 M NaCl. While, with increasing temperature, MnMo double oxide is gradually transformed into molybdate ions and permanganate [27]. Mn1 x Mox O_2 + x iron was added to enhance the stability of the catalyst at high temperatures. Subsequently, his Mn, Mo, and W tri-oxide catalysts containing an IrO_2/Ti support were reported by El Moneim et al. [28]. The electrical conductivity of the deposited oxides increased with the addition of tungsten, which also increased the OER efficiency of the anode. NiFeOx belongs to a

Table 4.1 Electrocatalysts with their reported OER performance

Catalyst	Electrodes	Overpotential [V]	Tafel Slope (mV/dec^{-1})	Current density mV cm^{-2}	Ref
Cobalt-iron layered double hydroxides	Glassy carbon	198 mv	39	10	[6]
LiC$_0$O$_2$	Glassy carbon	380	33.8	10	[7]
Iron–nickel	Glassy carbon	310	70.3	400	[8]
Co$_4$N	-	257	44	10	[9]
Mn-CuCO$_2$Se$_4$	NF	345	-	10	[10]
LiCO$_{0.8}$Fe$_{0.2}$O$_2$	Glassy carbon RDE	340	50	10	[11]
Ni–Fe oxide	GC electrode	328	42	10.2	[12]
LaC$_0$O$_3$-BaC$_0$O$_3$	GCE	266	129	10	[13]
LaMnO$_3$	GC electrode	324	74	10	[14]
Fe$_2$Se$_3$/Fe$_2$O$_3$	Ni Foam	160	30.2	20	[15]
Fe-doped C$_0$M$_0$O$_x$	GCE	290	42	10	[16]
Fes/Co$_3$S$_4$	-	252	58	10	[17]
(La$_2$NiO$_{4+\delta}$)	Glassy carbon rotating-disk electrode	-	50	1.6 V$_{RHE}$	[18]
FeC$_0$S$_y$/NCDS	GCE	284	52.1	10	[19]
CuSCS$_4$/CO$_0$	Ni-foam	179	46	10	[20]
Ni–Fe-V trimetallic phosphorus-selenium composite	Carbon cloth	168	44.32	10	[21]
NiCo-LDH and Co$_9$S$_8$	Rotating disk electrode	151	100	10	[22]
CoFe@PANI	Ni foam	237	46	10	[23]
Fe-Co$_3$O$_4$ NBs	-	265	54.2	10	[24]

different class of metal oxides with OER activity and frequently shows remarkable activity in basic conditions, where Fe inclusion is crucial for enhancing performance. OER kinetics enhancement by Fe doping jactitation, however, is still up for discussion. Highly active NiFeOx electrocatalysts eventually lose their activity due to Iron loss through active sites during catalysis. The CeOx layer's anodic deposition was done by Obata et al. to stop Fe loss to have high stability and catalytic activity. While allowing OH and O$_2$ removal through without disturbing the catalytic reaction, the CeOx layer effectively prevented redox ions from diffusing through the layer [29]. In basic simulated seawater, Pb 2Ru$_2$O$_7$x was found to be superior to neutral electro-catalysts of OER activity, selectivity [5], and stability RuO$_2$ was found to have much lower selectivity, activity, and stability than Pb$_2$Ru$_2$O$_7$x in basic and neutral conditions of simulated seawater. Ruthenium's high oxidation state and high O_2 density availability, $Pb_2Ru_2O_{7x}$ exhibited exceptional OER-selective electrocatalytic activity. Much work has been done, but in acidic media, he finds it very difficult to enhance the effectiveness of OER. Electrodeposited MnOx films by Koper et al. at iridium oxide-filled glassy carbon electrodes at 0.9 pH in chloride ions solution. MnOx deposition on IrOx decreased the CER selectivity from 86% to less than 7% in 30 mM Cl [30]. Metal hydroxide and oxides based electrocatalyst for oxygen evolution reactions are shown in Fig. 4.1.

An important issue with electrolysis in marine situations is the corrosion that chloride ions cause. Juodkazyte et al. suggested the formation of a nickel oxide layer on conductive glass in a chloride solution with remarkable corrosion resistance. Furthermore, Okada et al. produced dual-layer films having an upper Mg^{2+} layer intercalated Co and a lower Co (OH) layer. $CoMnO_2$ served as a barrier, allowing water to access the Co (OH) $_2$'s active catalytic sites while simultaneously ejecting chloride ions, even though it was not catalytically active [33]. In the $MgCl_2$ solution, Mg| CoMnO 2/Co (OH)$_2$ represents a prominent selection with maximum OER in the absence of ion exchange Metallic hydroxides are the catalyst often used for OER along with metal oxides and, in some cases, even outperformed IrO_2. However, due to the low conductivity and chloride corrosion, most hydroxide catalysts fall short of meeting the expanding industrial requirement of saltwater electrolysis. Layered double hydroxides are commonly discussed metallic hydroxide catalysts (LDHs). Recently, NiFe-LDHs with partially and completely crystalline topologies for effective OER efficiency were reported [34]. In OER testing, it was found that partly crystalline NiFeLDH outperformed its fully crystalline counterpart. An S-doped Ni/Fehydroxide catalyst (S (Fe, Ni) OOH) with outstanding OER activity in both seawater and alkaline electrolytes was disclosed by Yu et al. at current densities of 500 and 100 mA cm^2, the over-potentials of S(Ni, Fe)OOH in the saltwater containing 1 M KOH were 300 and 398 mV. Using Zr-doped CoFeLDH/NF, Liu et al. generated highly active OER catalysts [35]. Results showed that adding Zr^{4+} reduced crystallinity and increased active surface area. The surface produced COOH intermediate improved OER activity and prevented Cl from adsorbing. The electrocatalyst overpotential in a 1 M KOH + 0.5 M NaCl was only 303 mV when the current density reached 100 mA cm^2 [36].

4.2 Noble Metals Based Electrocatalysts

Effective electrode materials for OER have long been thought to be noble metal and metallic oxide electrocatalysts. Examples include cutting-edge electrocatalysts for OER, RuO_2 and IrO_2, frequently cited as examples. Nevertheless, the main issues are greater cost and severe solubility of RuO_2 and IrO_2, which draw a lot of attention to the need to modify the catalysts to permit chemistry and morphology tuning [37]. Numerous methods are suggested to increase electrocatalysts' activity and chemical stability and lower their greater cost. There is great interest in heteroatom doping as

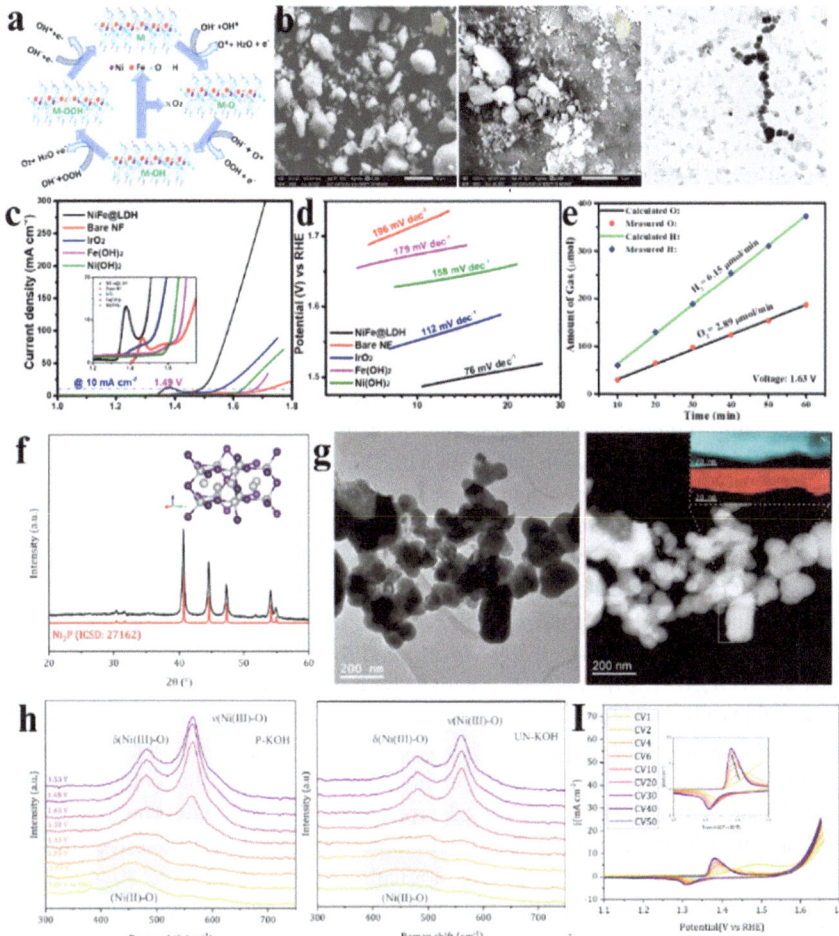

Fig. 4.1 (a) Plausible OER Mechanism of NiFe@LDH; (b) SEM and HR-TEM Image of NiFe@LDH lattice structure; (c) Polarization (OER) curves of NiFe@LDH/NF, IrO2/NF, Ni(OH)2/NF, Fe(OH)2/NF, and bare NF: OER LSVs; (d) slopes of Tafel; (e) Plot Faradaic efficiency Reused with permission from ref [31] copyright 2024 American Chemical Society.; (f) XRD pattern for Ni2P composite; (g) TEM images of Ni2P composite; (h) Raman spectra of activated Ni2P at different potentials in P-KOH and Un-KOH; (I) Fifty continuous CVs for the activation of Ni2P recorded at a scan rate of 20 mV s–1 in 1.0 M P-KOH Reused with permission from ref [32] copyright 2024 American Chemical Society

a means of fine-tuning the chemistry of IrO_2-based OER electro-catalysts. However, various guest atoms in the host system created their energy domains. Chen et al. prepared hollow porous polyhedral-shaped Cu-doped RuO_2 composed of very small nanocrystals. The catalyst exhibits excellent OER performance with less overvoltage of 188 mV at 10 mA cm^2 at low pH electrolyte and greater stability over 10,000 cycles of endurance testing [38]. According to high-resolution TEM and XRD, high-index facets are evidence that Cu was added into RuO_2 to produce the Cu-doped RuO_2 rutile phase. According to DFT calculations, the high OER activity is the presence of highly under-coordinated Ru sites (CN = 3) on the high refractive index surface, which can significantly reduce the OER overvoltage. There is a nature. RDS was found to be involved in the development of *OOH on RuO_2 (110), the energy barrier is 0.78 eV. 0.66 EV energy is required to break the barrier of other high-index facets on the RuO2 (111) area, reducing the 120 mV over-voltage. The Cu dopant RuO_2 modifies the electronic structure showing a wide binding region close to the p-band-centered Fermi level and can further enhance the OER activity by forming Ru unsaturation sites with O generated by the Cu dopant water surface.

OER catalyst development can successfully change the structure and maintain the adsorption energies of reaction intermediates by combining Ru or Ir with d-block metals. Using a eutectic-directed self-annealing approach, a group created by Zhang et al. of IrM (M = Fe, Co, Ni) catalysts with the structure of entangled nonporous nanowires [39]. The transition metal-dependent features are evident from the results. The IrNi NWs demonstrate the highest OER activity compared to the IrFe and IrCo NWs, and the lowest overvoltage at 10 mA cm^2 is 283 mV. The DFT calculation results explain the increased activity of the IrNi NWs. M and Ir oxidized at greater potential during the OER process to form IrMOx. IrO_2, IrFeOx, IrCoOx, and IrNiOx have d-band centers at 3.61, 3.72, 4.34 eV, and 4.09. The density of states shifted negatively (DOS). The d-band electron distribution far from the Fermi level induced by the effect of ligand after alloying is indicated by the less shift of the d-band center of Ir. OER activity is strongly influenced by the binding energies of OH, OOH, and O species To demonstrate the 3d transition metallic OER activity that could downshift the Ir d band center and decrease the adsorption capacity of the intermediates.

Modifying the structure is important to expose the catalytically active sites and benefit from interfacial effects. Morphological control has attracted attention as a surface modification. For example, hollow nanoparticles with enhanced catalytically active sites, such as nanoshells, nanocages, and nanoframes shown to be effective in enhancing catalysis. As reported by Lee et al., Ir-based multilayer metal bilayer nanoframe (DNF) electrocatalysts prepared by single-step synthesis was shown. TEM, HAADF-STEM, HRTEM, and atomic mapping shows the preferred IrNiCu-DNF (ternary alloy) retains the rhombic dodecahedral structure after strong acid etching exhibits even particle spreading throughout DNF. The framework structure of the IrNiCu catalyst for OER, which prevents particle coarsening and aggregation during the OER process and the in-situ fabrication of the enhanced rutile IrO_2 phase, is responsible for the good electro-catalysis and durability increase. It was also shown that the electrical conductivity of the catalyst was affected by morphology control, increasing the OER electrocatalytic activity. IrO2 nanoneedles were developed by Lee et al. have used a scalable molten salt method and demonstrated his OER activity and stability superior to those of IrO_2 nanoparticles [40]). comparing the OER electrochemical ability of IrO_2 and IrO_2 NPs nanoneedles, the IrO_2 nanoneedles possess higher stability and activity than the IrO_2 NPs. The ultrafine IrO_2 nanoneedles have a conductivity of 318.3 S.cm1, in contrast to the conductivity of 25.9 S cm1 for unshaped IrO_2 nanoparticles, making the geometry a highly active catalyst for electron transfer induced OER.

4.3 Metal Phosphides Based Electrocatalysts

Due to their excellent stability, conductivity, and metal coordination effects, metal phosphides have recently gained much attention [35]. Like hydrogenase, metal phosphides have a structure where the metal site and P site can act as the center of the hydrogenation receptor and the proton receptor, respectively. Wu et al. prepared Ni_2P-Fe_2P by dipping nickel foam in ferric nitrate and hydrochloric acid, then phosphating it [41]. Seawater electrolysis benefits from Ni_2P-excellent Fe_2P's activity because of its quick electron shifting rate, hydrophilicity, and resistance to corrosion. In seawater with a concentration of 1 M KOH, the overpotentials of Ni_2P-Fe_2P were 581 and 774 mV at current densities of 500 and 100 mA cm^2, respectively. Cobalt-doped Fe_2P (Co-Fe_2P) was used by Wang et al. through simulated seawater having 1.0 M KOH and 0.5 M NaCl, and it demonstrated excellent electrochemical durability of 22 h operation as well as an over-potential of 460 mV at 100 mA cm^2 [42]. Mechanism regarding metal phosphides based catalyst are shown in Fig. 4.2.

4.4 Metal Nitrides Based Electrocatalysts

The metal nitrides are attractive as an interstitial alloy in electrocatalysis because of their exceptional mechanical strength, excellent conductivity, strong stability, and superior corrosion resistance. Nitrogen atoms can be added to transition metals to change the density of states in the d band. The addition of nitrogen decreases the d-band density, resulting in greater catalysis than the parent metallic material. Ni_3N/Ni, NiMoN, and NiFeMoN are the most popular metal nitrides due to their excellent catalytic performance. Yu et al. made a three-dimensional core–shell NiMoN@NiFeN having NiFeN on NiMoN nanorods integrally adjusted on a nickel foam support to create a catalyst with a greater active site [46]. Natural seawater containing 1 M KOH, this catalyst exhibited overpotentials of 398 and 369 mV at current densities of 1000 and 500 mA cm^2, respectively. The anode surface underwent in situ development, and the resulting amorphous layers of NiFe hydroxide and NiFe oxide were the actual active sites. Moreover, the three-dimensional NiMoN@NiFeN core–shell features multi-layered pores, providing abundant active sites, enabling efficient charge transfer and rapid outgassing. Consequently, NiMoN@NiFeN showed outstanding OER performance and chlorine corrosion resistance. Additionally, a two-electrode seawater electrolyzer could be created by combining NiMoN and NiMoN@NiFeN as cathode and anode catalysts. The minimum voltage for this electrolyzer in naturally alkaline seawater at a current

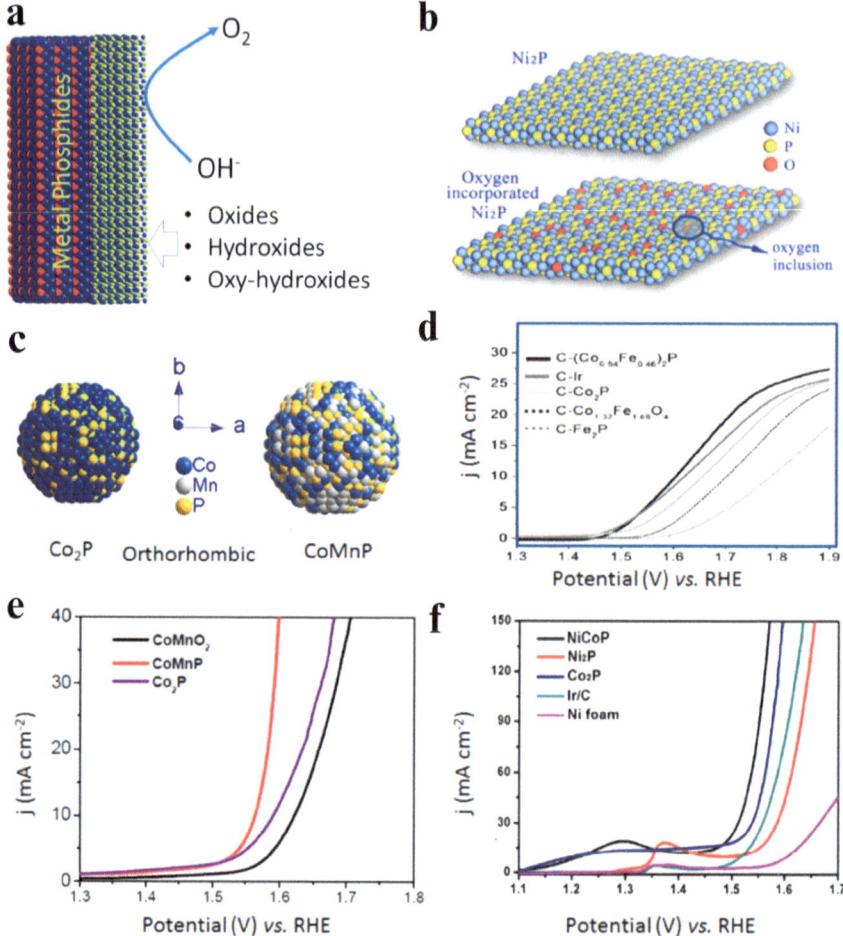

Fig. 4.2 (a) Sketch diagram of metal phosphides have emerged as efficient electrocatalysts for oxygen evolution reactions OER; (b) Schematic atomic models showing oxygen incorporation in Ni2P; (c) Atomic model showing Co_2P and CoMnP; (d) LSVs of composition variations of CoFeP in modified electrodes Reused with permission from ref [43] copyright 2015 Wiley; (e) LSVs of Co_2P-, CoMnP-, and $CoMnO_2$-modified electrodes [Reused with permission from ref [44] copyright 2016 American Chemical Society; (f) LSVs of NiCoP, Ni_2P, and Co_2P grown on Ni foam Reused with permission from ref [45] copyright 2015 Springer

density of 500 mA cm^2 was only 1.608 V. similar oxygen evolution mechanism is depicted in Fig. 4.3.

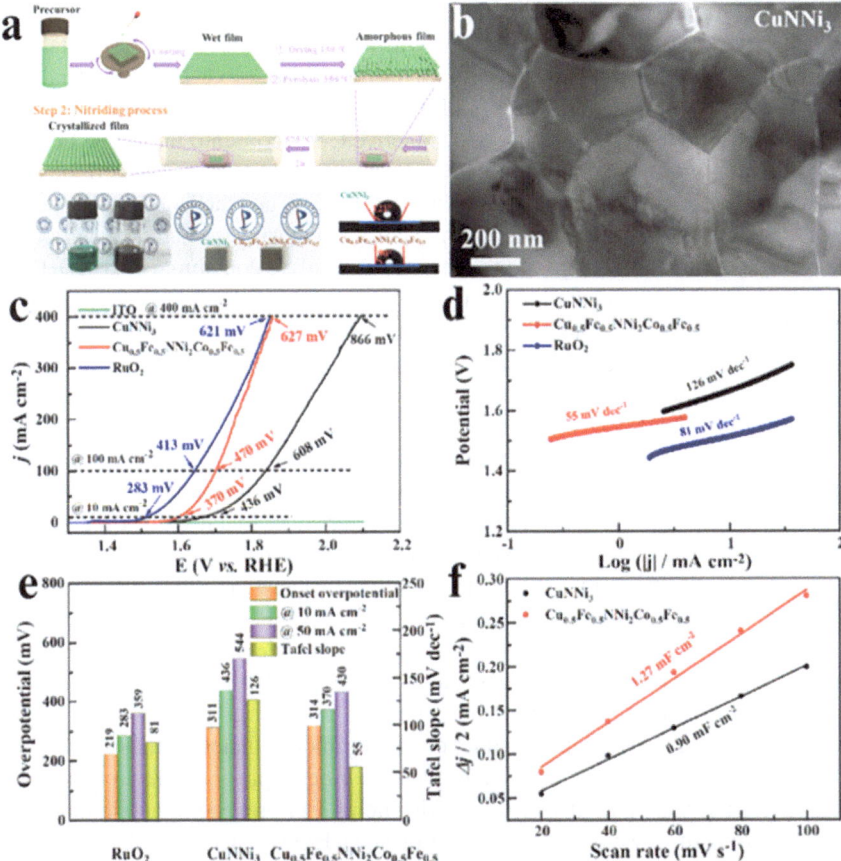

Fig. 4.3 (a) Schematic illustration of fabrication process for (Cu0.5Fe0.5)N(Ni3-2yCoyFey) (0 ≤ x ≤ 1); (b) SEM images; (c) LSV curves of electrocatalytical oxygen evolution test in 1.0 M KOH; (d) Tafel plots; (e) Comparisons of the ηonset, η10, η50 and Tafel slope; (f) Cdl results, Reused with permission from ref [47] copyright 2024 Elsevier

4.5 Hybrid Electrocatalysts

The studies have reported that catalysts with hybrid structures show remarkable activity for seawater electrolysis. However, in single metallic materials, the two components of hybrid catalysts can act synergistically, resulting in mixed charge distribution and increased electrocatalytic activity. Choi et al., For example, fabricated a heterostructure catalyst composed of Fe–Ni (OH)$_2$ nanosheets and Ni$_3$S$_2$ nanoarrays in an alkaline seawater electrolyte. He exposed many active sites in OER [48]. At a current density of 10 mA cm^2, the Tafel slope was only 46 mV dec.1, and the overpotential was as low as 269 mV. Moreover, with a faradaic efficiency of up to 95%, at a high current density of 100 mA cm^2, he shows remarkable stability after

27 h of operation. Iron activators can provide numerous heterostructures with active and selective sites to enhance the catalytic activity of FeNi $(OH)_2/Ni_3S_2$ catalysts. Furthermore, Wu et al. fabricated a core–shell structure of CoPx@FeOOH containing multiphase cobalt phosphide (CoPx, CoP-CoP$_2$) cores [49]. A synergistic effect was achieved by integrating an extremely conductive CoPx core into the active FeOOH shell. It features high conductivity, high surface area, and higher turnover frequency. The adsorption energies between the OER intermediates and the OOH active species were tuned by the negatively charged CoPx cores, increasing the catalytic activity of the catalysts. The overvoltage of the seawater electrolyte containing 1 M KOH at a current density of 100 mA cm^2 was only 283 mV and 337 mV at a current density of 500 mA cm^2. Exceptional stability without hypochlorite formation was also achieved over 80 h. Schematic of the mechanism for photocatalystic evolution reaction over MCS@a-Ni3S2 under visible light irradiation and other hybrid electrocatalyst are depicted in Fig. 4.4.

The metal (hydrogen) oxides, borides, phosphides, nitrides, and hybrid catalysts have all been successfully used in seawater electrolysis reactions. Each of these catalysts in seawater electrolysis has its characteristics. Metal oxides are usually cheap and very stable. Metal hydroxides are categorized by low cost, great catalytic proficiency, and a pronounced layered structure that increases the active surface. Metal hydroxides and oxides are the most common OER catalysts; however, their inherently low electrical conductivity is the main obstacle. Recently, metal nitrides have become one of the most studied OER catalysts due to their outstanding corrosion resistance, low resistivity, and high melting point.

Metal phosphides typically conduct electricity more efficiently than oxides and hydroxides while also being very stable. Metal borides typically exhibit outstanding catalytic activity and stability in OER. Also helpful in preventing anode materials from corroding in seawater is the metal boride layer. Hybrid catalysts, which combine different materials or structures, are gaining more and more attention as a way to utilize the benefits of various materials fully. Lattice oxygen species are indulged in the OER. In conventional AEM, the whole reaction occurs at a single metal site, with a minimum theoretical overpotential of 0.37 eV due to scaling relationships between OER intermediates. A new OER mechanism involving the lattice oxygen-mediated mechanism (LOM) and lattice oxygen species was recently proposed. With a lattice oxygen-mediated mechanism, the catalyst's lattice oxygen directly contributes to the oxygen evolution reaction. Recently, the involvement of lattice oxygen in catalytic oxidation reactions has been shown to occur in the gas phase of alloy catalysts. Interestingly, this phenomenon is thought to represent different reaction pathways, which may be the best suitable in OER electrocatalysis.

Indeed, the mechanisms involving lattice oxygen species have evolved thanks to several important studies. Based on the DFT study, Stevenson et al. suggested a first-order reaction pathway in which lattice oxygen participates in the OER reaction through the reversible development of surface oxygen vacancies. They presented a series of cobaltite perovskites and showed how the oxygen vacancies, metal–oxygen covalent bonds, and his OER activity are related. The correlation between the Co–O covalent bond and oxygen vacancy concentration is shown in DFT studies show

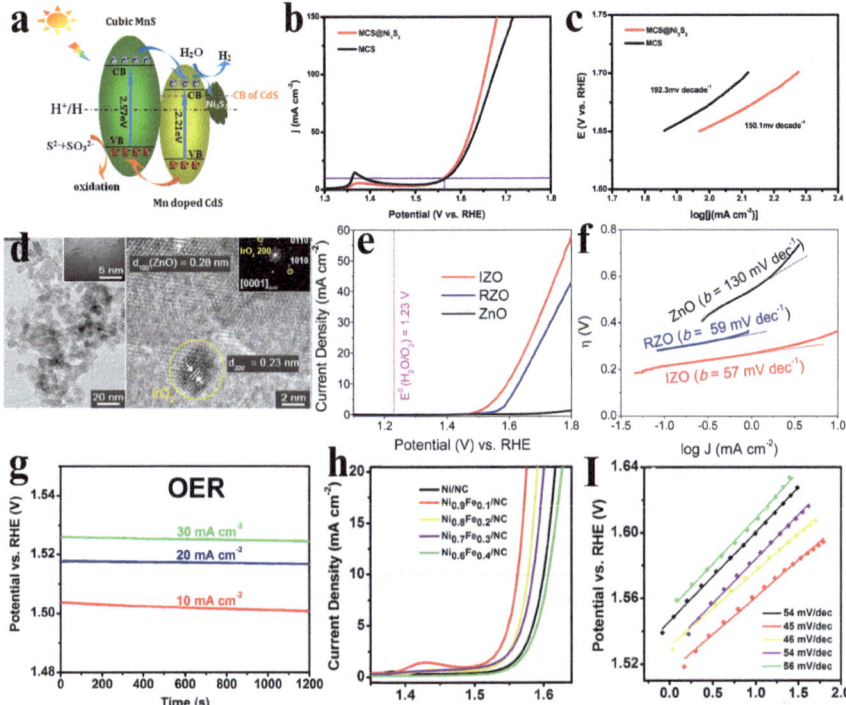

Fig. 4.4 (a) Schematic of the mechanism for photocatalystic evolution reaction over MCS@a-Ni3S2 under visible light irradiation; (b) LSV curves of OER for MCS and MCS@a-Ni3S2; (c) Tafel slopes of MCS and MCS@a-Ni3S2 Reused with permission from ref [50] copyright 2019 Elsevier (d) HRTEM and lattice-resolved and corresponding FFT images for IrO2–ZnO hybrid nanoparticles; (e) LSV curves of IZO, RZO, and ZnO for OER (in O2-saturated 1 M KOH) and (f) corresponding Tafel plots Reused with permission from ref [51] copyright 2017 American Chemical Society; (g) Loading of the catalysts on Ni foam at different constant current densities; (h), (I) Electrochemical water-splitting activities of $Ni_{1-x}Fe_x/NC$ on GC electrodes in 1 M KOH(aq) electrolyte LSV polarization curves for the OER and corresponding Tafel plots for the OER Reused with permission from ref [52] copyright 2016 American Chemical Society

that replacing Co^{3+} with Sr^{2+} brings the Fermi level (EF) quicker to the O2p band, supplemented by the increased overlap of the M 3d and O 2p bands, indicating stronger metal–oxygen binding. At the same time, ligand holes are generated simultaneously. The structural arrangement reduces the energy required to reach a steady state by causing oxygen vacancies when O_2 is formed and released. The low valence Sr^{2+} can be replaced by $La_{1x}Sr_xCoO_3$ (LSCO) structure to control the increase in vacancy concentration in the catalyst caused by the increase in covalent valence between metal–oxygen bonds. According to DFT modeling and experimental data, oxygen vacancy, OER activity, and oxygen diffusion rate are directly correlated this data demonstrates a correlation between increased OER activity, surface exchange kinetics, and increased vacancy. Oxygen in the lattice reacts with oxygen adsorbed on metal sites to form oxygen vacancies, which differs from the commonly proposed adsorbed -O species in the AEM mechanism. The relative stability of these two intermediates is important in determining whether OER occurs via her AEM or LOM for a given LSCO composition. A larger x is predicted to induce AEM to LOM transitions in the $La_{1-x}Sr_xCoO_{3-\delta}$ system, minimizing the energy of O vacancy formation and decreasing bulk stability. Stability and performance evaluation of hybrid electrocatalyst are shown in Fig. 4.5.

Shao Horn et al. gave the experimental indication that lattice oxygen is used to synthesize molecular oxygen in the OER using in situ ^{18}O isotope-labeled mass spectrometry [55]. They are $La_{0.5}Sr_{0.5}CoO_3$ and $SrCoO_3$. As a result, the increased covalent bonding between metal and Oxygen activates the OER mechanism. $La_{0.5}Sr_{0.5}CoO_3$, $Pr_{0.5}Ba_{0.5}CoO_3$, and $SrCoO_3$, ^{18}O-labeled Co-based perovskites with different metal–oxygen covalent valences, were used to determine the lattice oxygen content using on-line electrochemical mass spectrometry (OLEMS), which determines the contribution of Identify oxidation reactions. Mass spectrometry was used to measure oxygen gases of various molecular weights in situ, including $32O_2$ (16O16O), $36O_2$ (18O18O), and $34O_2$ (16O18O). Two separate lattice oxygen species are believed to be involved, according to the oxidation mechanism suggested for explaining the fabrication of $36O_2$(18O18O) and $34O_2$(16O18O). both chemical processes producing oxygen vacancy activity and molecular O_2 occurring at oxygenated surface sites were involved in these processes. Further, Kolpak et al. created active volcano maps for AEM and LOM. By reducing the thermodynamically required overvoltage, LOM shows higher OER activity than AEM [56]. AEM has a minimum theoretic overvoltage of 0.37 eV based on scaling relationships. For LOM, the relative constant G for VO + OO* VO + OH* is much lesser than the AEM-based G of 3.2 eV.

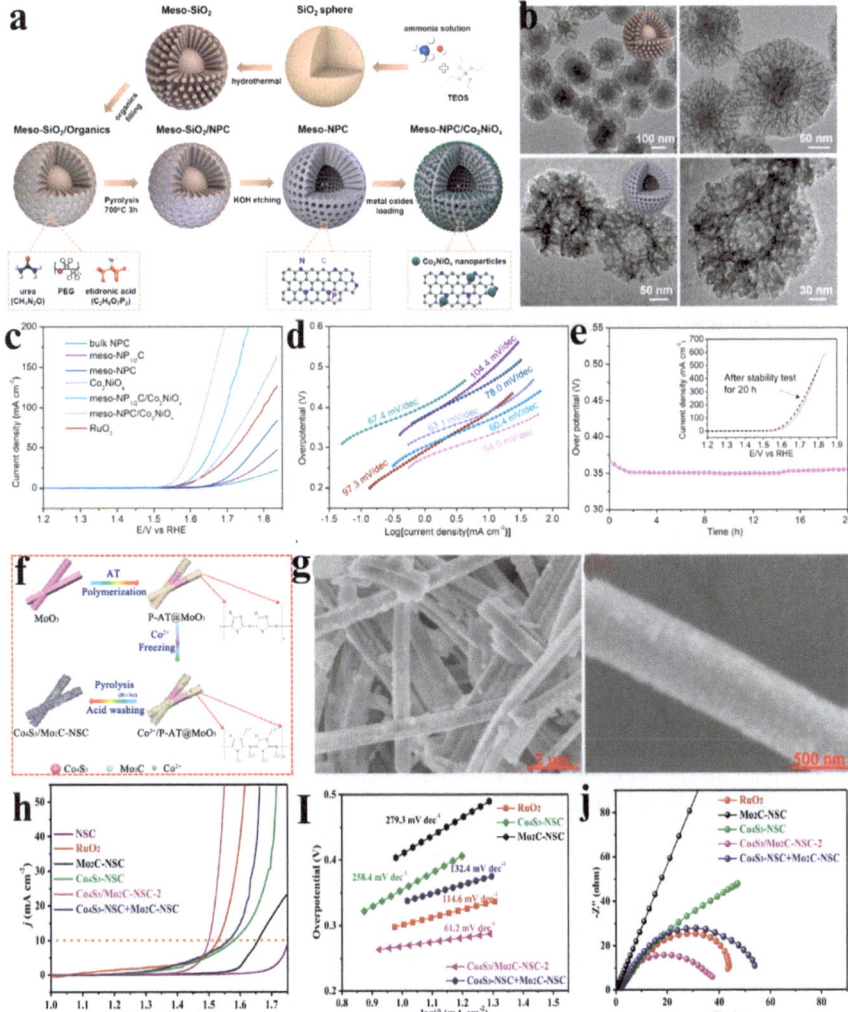

Fig. 4.5 (a) Schematic Illustration of the Preparation of the Meso-NPC/Co₂NiOₓ Hybrid Electrocatalyst for the OER Process; (b) TEM images of template meso-SiO₂; (c) Linear sweep voltammetry curves; (d) relevant Tafel plots of studied samples tested in 1.0 M KOH solution; (e) Chronopotentiometry curves at a constant current density of 10 mA cm–2 for the meso-NPC/Co2NiOx sample Reused with permission from ref [53] copyright 2020 American Chemical Society; (f) Schematic illustration of the synthesis routes of the Co₄S₃/Mo₂ C-N SC; (g) SEM images of the Co₄S₃/Mo₂C-NSC; (h) OER polarization curves of corresponding overpotential; (I) OER polarization curves of Co₄S₃/Mo₂C-NSC-2, Mo₂C-NSC, Co₄S₃-NSC, Co₄S₃-NSC + Mo₂C-NSC and RuO₂ in 1.0 M KOH solution at a scan rate of 5 mV s⁻¹, Nyquist plots (j) OER polarization curves for Co₄S₃/Mo₂C-NSC-2, Co₉S₈-NSC, Co-NSC, Mo₂C-NSC, NSC, and RuO₂. Reused with permission from ref [54] copyright 2024 Elsevier

References

1. Sohrabnejad-Eskan I et al (2017) Temperature-dependent kinetic studies of the chlorine evolution reaction over RuO2 (110) model electrodes. ACS Catal 7(4):2403–2411
2. Kraft A et al (1999) Electrochemical water disinfection part I: hypochlorite production from very dilute chloride solutions. J Appl Electrochem 29(7):859–866
3. Vos J, Koper M (2018) Measurement of competition between oxygen evolution and chlorine evolution using rotating ring-disk electrode voltammetry. J Electroanal Chem 819:260–268
4. Juodkazytė J et al (2019) Electrolytic splitting of saline water: durable nickel oxide anode for selective oxygen evolution. Int J Hydrogen Energy 44(12):5929–5939
5. Gayen P, Saha S, Ramani V (2020) Selective seawater splitting using pyrochlore electrocatalyst. ACS Appl Energy Mater 3(4):3978–3983
6. Li P et al (2019) Boosting oxygen evolution of single-atomic ruthenium through electronic coupling with cobalt-iron layered double hydroxides. Nat Commun 10(1):1711
7. Zheng X et al (2019) Electronic structure engineering of LiCoO2 toward enhanced oxygen electrocatalysis. Adv Energy Mater 9(16):1803482
8. Dastafkan K et al (2020) Efficient oxygen evolution and gas bubble release achieved by a low gas bubble adhesive iron–nickel vanadate electrocatalyst. Small 16(32):2002412
9. Chen P et al (2015) Metallic Co4N porous nanowire arrays activated by surface oxidation as electrocatalysts for the oxygen evolution reaction. Angew Chem 127(49):14923–14927
10. Yu F et al (2022) Valence-modified selenospinels as ampere-current-bearing oxygen evolution catalysts. Appl Catal B 316:121649
11. Zhu Y et al (2015) A high-performance electrocatalyst for oxygen evolution reaction: $LiCo_{0.8}Fe_{0.2}O_2$. Adv Mater 27(44):7150–7155
12. Qi J et al (2015) Porous nickel–iron oxide as a highly efficient electrocatalyst for oxygen evolution reaction. Adv Sci 2(10):1500199
13. Mahmoudi E et al (2023) LaCoO3-BaCoO3 porous composites as efficient electrocatalyst for oxygen evolution reaction. Chem Eng J 473:144829
14. Li Q et al (2021) Phase engineering of atomically thin perovskite oxide for highly active oxygen evolution. Adv Func Mater 31(38):2102002
15. Sohail M et al (2024) An efficient Fe2Se3/Fe2O3 heterostructure electrocatalyst for oxygen evolution reaction. Int J Hydrogen Energy 52:1290–1297
16. Van Dang C et al (2024) Heterostructure of fe-doped CoMoO x/CoMoO x as an efficient electrocatalyst for oxygen evolution reaction. ACS Appl Mater Interfaces 16(8):9989–9998
17. Deng S et al (2024) Flower-like iron sulfide/cobaltous sulfide heterostructure as advanced electrocatalyst for oxygen evolution reaction. Int J Hydrogen Energy 51:550–557
18. Wu Y-H et al (2024) Probing surface transformations of lanthanum nickelate electrocatalysts during oxygen evolution reaction. J Am Chem Soc 146(17):11887–11896
19. Wu L et al (2024) Nitrogen-Doped carbon dots modified Fe–Co sulfide nanosheets as high-efficiency electrocatalysts toward oxygen evolution reaction. Small 20(4):2305965
20. Alothman AA et al (2024) Facile fabrication of CuScS2/CoO as an efficient electrocatalyst for oxygen evolution reaction and water treatment process. Int J Hydrogen Energy 49:564–579
21. Qin Y et al (2024) A Ni-Fe-V trimetallic phosphorus-selenium composite supported on carbon cloth as freestanding electrocatalyst for oxygen evolution reaction. Fuel 357:129857
22. Xie A et al (2024) NiCo-layered double hydroxides/Co9S8 with heterogeneous structure as ultra-high performance electrocatalyst for oxygen evolution reaction. Int J Hydrogen Energy 51:349–361
23. Rajpure MM, Jadhav HS, Kim H (2024) Layer interfacing strategy to derive free standing CoFe@ PANI bifunctional electrocatalyst towards oxygen evolution reaction and methanol oxidation reaction. J Colloid Interface Sci 653:949–959
24. Zhao D et al (2024) Preparing iron oxide clusters surface modified Co3O4 nanoboxes by chemical vapor deposition as an efficient electrocatalyst for oxygen evolution reaction. Energy Storage Mater 66:103236

25. Izumiya K et al (1998) Anodically deposited manganese oxide and manganese–tungsten oxide electrodes for oxygen evolution from seawater. Electrochim Acta 43(21–22):3303–3312

26. Fujimura K et al (1999) Oxygen evolution on manganese–molybdenum oxide anodes in seawater electrolysis. Mater Sci Eng, A 267(2):254–259

27. Ghany NA et al (2002) Oxygen evolution anodes composed of anodically deposited Mn–Mo–Fe oxides for seawater electrolysis. Electrochim Acta 48(1):21–28

28. El-Moneim AA et al (2010) Mn-Mo-Sn oxide anodes for oxygen evolution in seawater electrolysis for hydrogen production. ECS Trans 25(40):127

29. Wu X et al (2019) CeO x-Decorated hierarchical $NiCo_2S_4$ hollow nanotubes arrays for enhanced oxygen evolution reaction electrocatalysis. ACS Appl Mater Interfaces 11(43):39841–39847

30. Vos JG et al (2018) MnOx/IrOx as selective oxygen evolution electrocatalyst in acidic chloride solution. J Am Chem Soc 140(32):10270–10281

31. Sreenivasulu M et al (2024) Rational designing of nickel-iron containing layered double hydroxide [NiFe@LDH] electrocatalysts for effective water splitting. Energy Fuels 38(14):12888–12899

32. El-Refaei SM et al (2024) Ni-Xides (B, S, and P) for alkaline OER: shedding light on reconstruction processes and interplay with incidental fe impurities as synergistic activity drivers. ACS Applied Energy Materials 7(4):1369–1381

33. Okada T et al (2020) A bilayer structure composed of Mgǀ Co-MnO2 deposited on a Co (OH) 2 film to realize selective oxygen evolution from chloride-containing water. Langmuir 36(19):5227–5235

34. Görlin M et al (2017) Tracking catalyst redox states and reaction dynamics in Ni–Fe oxyhydroxide oxygen evolution reaction electrocatalysts: the role of catalyst support and electrolyte pH. J Am Chem Soc 139(5):2070–2082

35. Yu F et al (2018) High-performance bifunctional porous non-noble metal phosphide catalyst for overall water splitting. Nat Commun 9(1):1–9

36. Liu W et al (2021) Zr-doped CoFe-layered double hydroxides for highly efficient seawater electrolysis. J Colloid Interface Sci 604:767–775

37. Shi Q et al (2019) Robust noble metal-based electrocatalysts for oxygen evolution reaction. Chem Soc Rev 48(12):3181–3192

38. Su J et al (2018) Assembling ultrasmall copper-doped ruthenium oxide nanocrystals into hollow porous polyhedra: highly robust electrocatalysts for oxygen evolution in acidic media. Adv Mater 30(29):1801351

39. Wang Y et al (2019) Nanoporous iridium-based alloy nanowires as highly efficient electrocatalysts toward acidic oxygen evolution reaction. ACS Appl Mater Interfaces 11(43):39728–39736

40. Park J et al (2017) Iridium-based multimetallic nanoframe@ nanoframe structure: an efficient and robust electrocatalyst toward oxygen evolution reaction. ACS Nano 11(6):5500–5509

41. Wu L et al (2021) Heterogeneous bimetallic phosphide Ni2P-Fe2P as an efficient bifunctional catalyst for water/seawater splitting. Adv Func Mater 31(1):2006484

42. Wang S et al (2021) Synthesis of 3D heterostructure Co-doped Fe2P electrocatalyst for overall seawater electrolysis. Appl Catal B 297:120386

43. Mendoza-Garcia A et al (2015) Controlled anisotropic growth of Co-Fe-P from Co-Fe-O nanoparticles. Angew Chem Int Ed Engl 54(33):9642–9645

44. Li, D., et al., Efficient water oxidation using CoMnP nanoparticles. J Am Chem Soc 138(12):4006–4009

45. Li Y et al (2016) Ternary NiCoP nanosheet arrays: an excellent bifunctional catalyst for alkaline overall water splitting. Nano Res 9:2251–2259

46. Yu L et al (2019) Non-noble metal-nitride based electrocatalysts for high-performance alkaline seawater electrolysis. Nat Commun 10(1):1–10

47. Zhu L et al (2024) Design of high-entropy antiperovskite metal nitrides as highly efficient electrocatalysts for oxygen evolution reaction. Int J Hydrogen Energy 51:638–647

48. Cui B et al (2021) Heterogeneous lamellar-edged Fe-Ni (OH) 2/Ni3S2 nanoarray for efficient and stable seawater oxidation. Nano Res 14(4):1149–1155

49. Wu L et al (2021) Rational design of core-shell-structured CoPx@ FeOOH for efficient seawater electrolysis. Appl Catal B 294:120256
50. Li Z et al (2019) Mn-Cd-S@amorphous-Ni3S2 hybrid catalyst with enhanced photocatalytic property for hydrogen production and electrocatalytic OER. Appl Surf Sci 491:799–806
51. Kwak I et al (2017) IrO2–ZnO hybrid nanoparticles as highly efficient trifunctional electrocatalysts. J Phys Chem C 121(27):14899–14906
52. Zhang X et al (2016) Facile synthesis of nickel–iron/nanocarbon hybrids as advanced electrocatalysts for efficient water splitting. ACS Catalysis 6(2): 580–588
53. Wang J, Zeng HC (2020) Hybrid OER electrocatalyst combining mesoporous hollow spheres of N, P-Doped carbon with ultrafine Co2NiOx. ACS Appl Mater Interfaces 12(45):50324–50332
54. Liu Y et al (2020) A modulated electronic state strategy designed to integrate active HER and OER components as hybrid heterostructures for efficient overall water splitting. Appl Catal B 260:118197
55. Alsaç EP et al (2021) Structure–property correlations for analysis of heterogeneous electrocatalysts. Chemical Physics Reviews 2(3):031306
56. Yoo JS et al (2018) Role of lattice oxygen participation in understanding trends in the oxygen evolution reaction on perovskites. ACS Catal 8(5):4628–4636

Chapter 5
Electrocatalyst Design for Chlorine Evolution Reaction

Abstract The electrochemical chlorine evolution reaction (CER) is elementary reaction and commonly used for the conversion of Cl^- to Cl_2 in a seawater environment. Various precious metal based electrocatalyst and different design strategies have been utilized for engineering of CER electrocatalyst but most of the proposed materials fundamentally suffer from a selectivity problem between the CER and oxygen evolution reaction (OER). Hence, research and development of selective CER electrocatalysts vitally necessary. In this perspective, in this chapter, some advances CER electrocatalysts with superior performance was outlined.

5.1 Introduction

Combining electrochemical and spectroscopic techniques allowed researchers to examine the Cl_2 evolution reaction's (CER) reaction pathway. It is demonstrated that the interaction between RuO_2 and water leads to the oxidation and regeneration of the catalyst surface during CER [1]. The electrochemical chlorine evolution reaction (CER) is a crucial anodic reaction in the Chlor-alkali process for Cl_2 production, on-site formation of ClO^-, and Cl_2-mediated electrosynthesis. The Chlor-alkali industry and electrochemical wastewater treatment are two practical applications for the chlorine evolution reaction (CER). For energy-efficient chlorine generation, water disinfection, and wastewater treatment, efficient, stable, and affordable electrodes are essential. These real-world applications necessitate a thorough comprehension of the catalytic process of chlorine evolution, the requirement for less expensive electrode materials, and advancements in electrode design at the atomic level [1].

Chloride ions undergo chlorine evolution reactions (CER), which produce chlorine gas (Cl_2). Like this, protons are also released when water is oxidized into oxygen (O_2). The actions take place through.

$$2H_2O \rightarrow O_2 + 4H^+ + 4e^-$$
(5.1)

$$2Cl^- = Cl_2 + 2e^-$$
(5.2)

Z. K. Ghouri et al., *Hydrogen Production from Seawater Electrolysis*,
SpringerBriefs in Energy, https://doi.org/10.1007/978-3-031-73442-7_5

The normal hydrogen electrode (NHE) scale shows that the equilibrium potential CER is pH-dependent, but on the reversible hydrogen electrode (RHE) scale, it is pH-independent [2]. The main step in the Chlor-alkali process is CER, which facilitates the formation of bulk chemicals such as caustic soda and chlorine, which are essential in huge volumes throughout the chemical industry. The Chlor-alkali process uses a lot of energy while having very quick kinetics due to the 2e⁻ nature of the reaction and a too-high equilibrium potential. The power supply needed to cause chlorine development is the biggest commercial rate. Since the CER process is so large, much study has been done to identify the appropriate CER process parameters because even little efficiency improvements can have a big impact as oxygen is eco-friendly and can be released safely into the atmosphere, making OER the ideal anodic reaction in water electrolysis. This device's intended use is to finish the electrical circuit in water electrolysis cells, permitting the storage and collection of electrical energy in the form of hydrogen as a chemical fuel, thereby decreasing the unpredictable nature of renewable energy fluxes. The kinetics of Oxygen evolution reaction (OER) is very slow due to its four-electron nature, especially in contrast to CER. The challenge of initiating this reaction is one of the major obstacles impeding extensive use of water electrolysis for storing energy. It is promising to efficiently release chlorine with no anodic breakdown of the aqueous solvent in acidic aqueous environments with a pH less than 2 due to the simple CER kinetics relative to OER [3]. *CER and OER activity on iridium-based double perovskites shown in* Fig. 5.1.

Practically speaking, it is challenging to envision a scenario in which the combined development of chlorine and oxygen is desirable. The presence of OER is undesirable in Chlor-alkali because the oxygen poses an issue of safety, and its creation is connected to catalyst deterioration, in contrast to chlorine, a result of water electrolysis that is undesirable due to its environmental toxicity. As an OER-selective anode will enable the direct breakdown of saline water without the expensive requirement of eliminating chloride from the system, researching the selection between these two related reactions is crucial. Although OER and CER initially appear to be fundamentally different reactions, it has frequently been noted that effective catalysts at OER

Fig. 5.1 CER and OER activity on iridium-based double perovskites

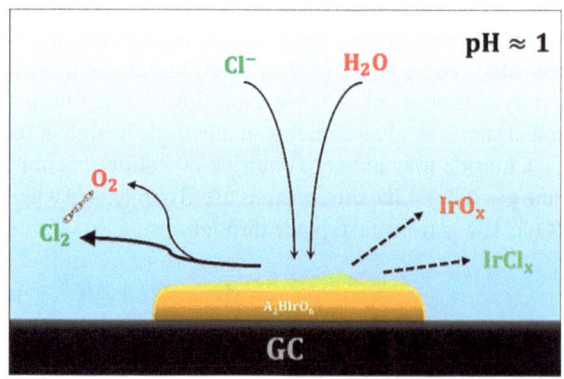

are also active for chlorine evolution reactions due to a so-called scaling connection amongst the binding energies of their significant intermediates.

This connection suggests that CER and OER are connected kinetically at an ultimate level, possibly through a shared active site or intermediate surface species. It also suggests that it may be difficult to maintain control over the degree of selectivity between the 2 reactions. Separating the two processes effectively may be difficult, if not impossible, based on kinetic factors, such as choosing an appropriate catalyst.

On metal oxides, both CER and OER can be catalyzed most easily. Due to thermodynamic limitations, CER on so-called dimensionally stable anodes (DSA) made up of combinations of TiO_2 and RuO_2 is often performed in acidic solutions. OER may be supported over an extensive range of pH; however, in terms of their action, electrolyzers named polymer electrolyte membrane (PEM) that use acidic pH and are best suited with Ir-based mixed metal oxides are state of the art. The only substance known to maintain long-term stability during acidic OER operation is iridium, either in its pure oxide form or as a dopant. Iridium is present in relatively few acidic OER anodes. Unfortunately, the absence of Ir substantially limits the widespread use of PEM electrolyzers. Scientific research has focused heavily on ways to lower the required Ir loading, such as increasing the active surface area or creating Ir-based materials with higher intrinsic activity. The Chlor-alkali process, among others, makes the chlorine evolution reaction (CER) a crucial reaction in electrochemistry. In regenerative fuel cells, precious metal catalysts are frequently employed to speed up the chlorine reaction. Furthermore, substantial quantities of precious metals are not required to serve as effective electrocatalysts for hydrogen-chlorine regeneration cells. Platinum and ruthenium oxide (RuO_2) are two of the maximum utilized electrocatalysts for the chlorine electrode [4].

Fundamental research currently mostly relies on computational simulations. The structure of metal or carbon materials can be affected by dynamic surface evolution processes such as the rearrangement of atoms, composition dissolution, internal porosity structure redistribution, and the high overpotential and strongly acidic environment. More experimental data are therefore required to determine the structure of the actual active species and subsequently to suggest a potential reaction mechanism using computational simulations. For this purpose, stretched X-ray absorption fine structure (EXAFS) and XANES analysis of X-ray absorption spectra (XAS) are necessary to identify the electronic structures and geometry of catalysts. To understand the catalytic mechanisms and potential synergistic effects, in situ Raman spectroscopy and FTIR experiments are also strongly advised for discovering intermediates throughout the process of chlorine oxidation. The future experimental design could be more effectively guided by results from computational models combined with cutting-edge methods like operando XAS or in situ HRTEM.

Although the CER is a straightforward two-electron process, its exact mechanism is still debatable. The basic catalytic pathways of the RuO_2 (110) demonstrated the role in the CER process and were investigated using various methods, including ab initio thermodynamic analyses and first-principles calculations. To validate theoretical predictions, extensive experimentation has been done in addition to theoretical calculations. Three mechanistic reaction routes have been proposed thus far: Volmer-Heyrovsky (V-H), Krishtalik, and Volmer-Tafel (V-T).

Volmer–Tafel

$$2Cl^- + 2^* \rightarrow 2Cl^* + 2e^-$$

$$Cl^* + Cl^* \rightarrow 2^* + Cl_2$$

Volmer–Heyrovsky

$$2Cl^- + {}^* \rightarrow Cl^* + e^- + Cl^-$$

$$Cl^* + Cl^- \rightarrow {}^* + Cl_2 + e^-$$

Krishatalik

$$2Cl^- + {}^* \rightarrow Cl^* + e^- + Cl^-$$

$$Cl^* + Cl^- \rightarrow Cl^{*+} + e^- + Cl^-$$

$$Cl^{*+} + Cl^- \rightarrow {}^* + Cl_2$$

* These are the live websites (surface oxygen or metal atoms). All three processes begin with a Cl adsorption procedure, as demonstrated in Eq. (5.1) through (5.3). At that time, three diverse processes occurred: (1) rearrangement of desorption as gaseous chlorine (Tafel) and two adsorbed chlorine sites; (2) direct recombination of one more Chlorine on the preexisting desorption as gaseous chlorine and adsorbed chlorine site(Heyrovsky); and (3) formation of chloronium ion, which afterward recombines with desorption and another Cl as gaseous chlorine (the process of Krishtalik). The V-H pathway was determined to be the best possible reaction pathway on the surface of RuO_2 (110) and most of the single-crystalline transition metal oxides by analyzing the first-principles research kinetic data. However, the experimental kinetic results alone could not support the actual catalytic process. To more extensively explore the mechanism of the CER electrochemical reaction, subsequent theoretical investigations used thermodynamic simulations. The unsaturated Rucus sites and Rubr sites are often shielded by the oxygen on the surface under the equilibrium potential since it was discovered that the RuO_2 (110) surface is completely covered with on-top oxygen.

Then, using this totally O-covered RuO_2 (110) surface as a foundation, Exner et al. applied density functional theory (DFT) to more precisely determine each step's loss of Gibbs free energy. how the Volmer step, with no theoretical loss of energy in the

Heyrovsky step and which has a 0.13 eV energy barrier, rates the V-H step at $U = 1.36$ eV.

The energy loss in the Krishtalik and Tafel phases, which is significantly more than that in the Volmer step, is around 0.98 eV and 0.36 eV, respectively. Therefore, among the three suggested paths, the Volmer-Heyrovsky (V-H) mechanism is the one that best explains the extraordinary catalytic capacity of the RuO_2 (110) surface. In addition, it was discovered that the RuO_2 activity could be further increased by lowering the Gibbs energy of the Volmer step by replacing the fractional top Ru sites with other noble atoms. This was founded on the Gibbs energy diagram of the V-H mechanism that is now in use. For instance, the Gibbs energy loss for the V-H process is reduced by 0.05 eV when a monolayer PtO_2 is doped on the RuO_2 (110) surface. Figure 5.2 depicts the schematic illustration of the electromigration, enrichment, and transformation of ions near the surface of nanotips.

Hanson et al. created a Pourbaix diagram to examine the stable surface structure as a function of supplied potential and pH better to understand the chlorine oxidation mechanism [2]. Three intermediates—ClOc, Cl(Oc)$_2$, and Clc—are depicted in the displayed Pourbaix diagram as the active species on the surface of RuO_2 (110) and help out in the evolution of chlorine as given in Eq. (5.4) through (5.6), respectively.

$$Oc + 2Cl^- \rightarrow ClOc + Cl^- + e^- \rightarrow Oc + Cl_2 + 2e^- \qquad (5.3)$$

$$Occ_2 + 2Cl^- \rightarrow Cl(Oc)_2 + Cl^- + e^- \rightarrow Oc + Cl_2 + 2e^- \qquad (5.4)$$

$$2Cl^- + c \rightarrow Cl^- + Clc + e^- \rightarrow Cl_2 + 2e^- \qquad (5.5)$$

Out of the three suggested intermediates, it is expected that process (5)'s O cc2 will most likely replace process (4)'s on-top oxygen Oc over a wider pH range in light of oxygen dissociation. The electrolyte's Cl will likely be adsorbed on the O cc2 to create Cl(Oc)$_2$ and then desorb as gaseous chlorine increases as the applied potential rises. Because Cl(Oc)$_2$ has a higher positive charge than O cc2, the catalytic pathway on the RuO_2 (110) facet is shown to be more akin to the mechanism of Krishtalik. A contrary theory was then put up, claiming that the intermediate Cl(Oc)2 is insufficiently stable and that the more thermodynamically stable ClOc is created instead when solvent effects on the surface being formed are taken into account (Fig. 2f). As a result, reaction route (4), which is more similar to the Volmer-Heyrovsky pathway, is still regarded as the primary mechanical procedure for the RuO_2 (110) Oxygen-covered surface. ClOc intermediate production is confirmed to be an important factor in CER by recent work that involved calculating the Gibbs energies of intermediates on RuO_2 and IrO_2 surfaces.

The unsaturated metal cus (c), along with the on-top oxygen active sites Oc, could function as a useful site for CER through the suggested method (6). However, active sites on the surface of RuO_2 are more numerous than active unsaturated metal cus sites (Rucus) (110). The surface completely covered in oxygen is more stable thermodynamically (Fig. 22 h). Since Oot invariably occupies the Rucus sites at

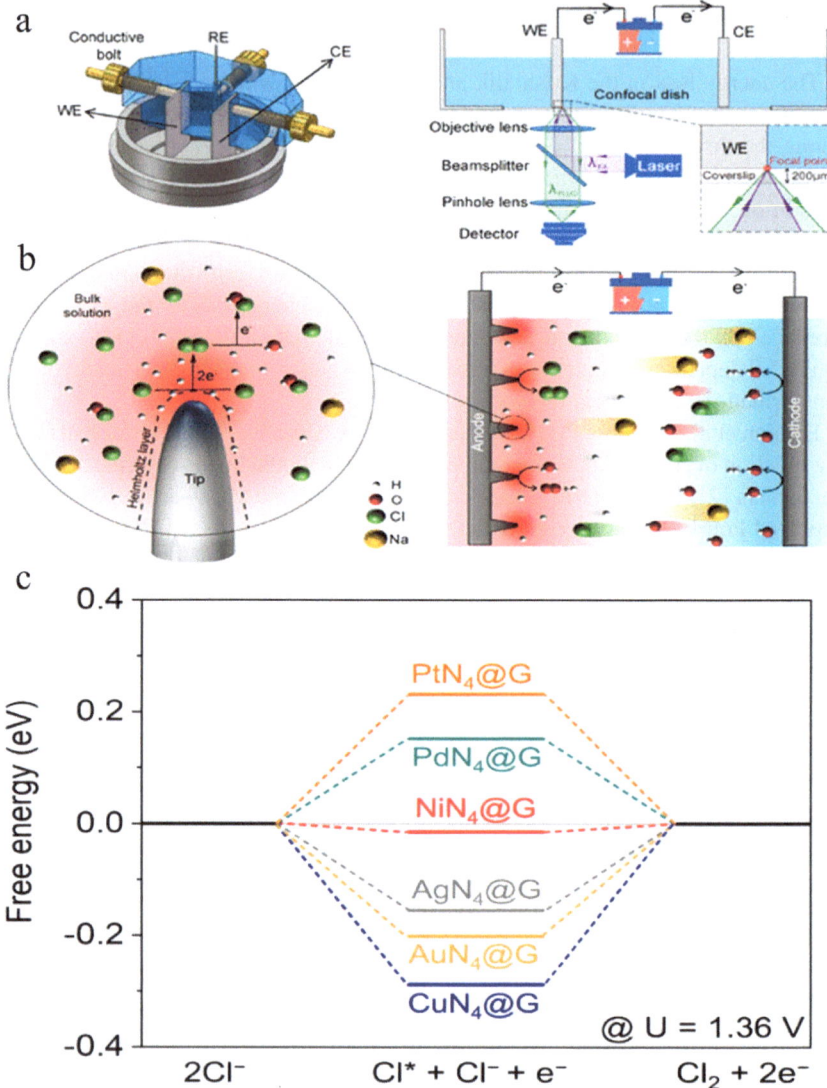

Fig. 5.2 (a) Schematic of the combined optical imaging and EC reaction setup (b) Schematic Illustration of the electromigration, enrichment, and transformation of ions near the surface of nanotips reused with permission from ref [5] Copyright 2022 American Chemical Society. (c) Gibbs free energy diagram for the CER over $TMN_4@G$ systems at the equilibrium potential of 1.36 V vs. SHE Reused with permission from ref [6] Copyright 2022 Elsevier

Table 5.1 The electrocatalysts reported for Chlorine oxidation reaction (CER)

Catalyst	Electrodes	Overpotential [V]	Tafel Slope [mV/dec]	Current Density	Ref
Pt1/CNT	KCl-saturated	30	38	10	[7]
Pt1/CNT	Ag/AgCl	40	40	10	[8]
Pt-RuO$_2$	SCE		101.7	80	[9]
Pt-SA/NiV-LDH	Hg/HgO	130	50	39	[10]
Ru–S-TiO$_2$	Pt mesh	5	67	10	[11]
RuSnTi	SCE	1.15	40	96.6	[12]
Ru$_x$Co$_{3-x}$O$_4$	Ti mesh	10	45	500	[13]
Ru–Cl–N SAC	GC	12	23.9	10	[14]
Ru, W–NiSe$_2$	NiSe2/NF	100	133	10	[15]
Ru-NiMoO$_4$/Ti	Ti foam	20	44.8	10	[16]
CoNi@MoS$_2$–PdSARuSA	CoNi@MoS$_2$	89	51	10	[17]
Ru/NiFeOOH/NFF	Nife foam	252	46.8	10	[18]
RuO$_2$/TiO$_2$	Ti	0.18	55	10	[6]
IrO$_2$/TiO$_2$	Ti	0.20	60	10	[19]
Co$_3$O$_4$/Ni	Ni	0.2	70	10	[20]

equilibrium, they are hard to depict and cannot be used to adsorb Chlorine from the electrolyte directly. The Pourbaix diagram indicates that the V-H mechanism is the preferred pathway for chlorine oxidation; however, all identified paths are technically possible on the RuO$_2$ (110) surface [2].

Several electrocatalysis has been reported for their best catalytic efficiencies in CER. Some of them are given in table 5.1 below (Table 5.1).

5.2 Platinum Based Electrocatalysts

The Pt electrocatalyst, especially as nanoparticles on a carbon substrate (Pt/C), demonstrates superior electrochemical performance for the chlorine reduction reaction (CRR), but it achieves worse electro catalytically for the chlorine evolution reaction (CER) [21]. In fuel cells, as a result of its superior stability and increased catalytic activity compared to other noble metals, platinum is one of the greatest popular metals utilized as a catalyst. The effect of a Pt electrode on both chlorine development and its reduction was initially studied in 1935 by Chang and Wick [22]. They demonstrated that, in line with Erdey-Gruz and Volmer's idea, this electrode performs best for chlorine evolution rather than reduction. Yeo et al. examined the reaction of a hydrogen-chlorine regeneration cell utilizing various chlorine-sided electrodes over the ensuing few years [23]. Pt was one of the catalysts tested; however, it was confirmed that the electrode corroded because chlorine's potential (0.27 V) was higher than that of Pt/PtCl$_4$. Because of this, the cell's mass transfer was restricted after discharge. Thomassen et al. assessed the efficiency of the fuel cell system of hydrogen-chlorine employing gas diffusion electrodes at the start of the twenty-first century. They also looked at Pt's durability as an electrocatalyst for chlorine reduction. They concluded that a formulation with 20 weight percent platinum on carbon was unstable as a cathode material because platinum decomposed by adding chlorine and formed chloroplatinic acid. Using only platinum as a catalyst resulted in a 45% reduction in cell performance.

It was recently discovered that atomically scattered catalysts increase the precious metals' catalytic efficiency. Specifically, it has been shown that a carbon nanotube electrocatalyst with atomically scattered Pt-N4 sites (Pt1/CNT) is excellently active and selective in catalyzing CER. Tafel investigations revealed that Pt1/CNT had a faster rate of CER kinetics (38 mV dec1) than more traditional catalysts like Ru/Ir and PtNP/CNT built on. Thus, it may be deduced that it is not as practical to use simply Pt electrode as an electrocatalyst where the oxidation–reduction reaction of chlorine occurs as it appeared in the initial investigation. Due to its poor electrocatalytic activity and stability for CER, platinum is prohibitively expensive. Given that the chief concern material is carbon, which is prone to corrosion, the maximum practical choice is to enhance the structure of Pt as an electrocatalyst by adding additional metals, thereby adjusting materials of support. The efficiency of Pt as a catalyst during electrocatalytic oxidation may decrease because of the growth of intermediates that fill the active sites [24].

To increase Pt's catalytic activity, other metals are added. In 1992, Shibli et al. assessed the hydrogen-chlorine fuel cells' performance with several bimetallic catalysts, including chlorine reduction and hydrogen oxidation. They used Pt-Ru- (5 wt% Ru and 2.5 wt% Pt), Pt-Ir-loaded (5 wt% Ir and 2.5 wt% Pt), and Pt- (5 wt %) catalysts on electrodes of carbon gas diffusion to study the effectiveness of CRR in this system. The key finding of this study was that the CRR was improved by utilizing bimetallic catalysts, namely Pt-Ir. Furthermore, the stability of the system of the hydrogen-chlorine fuel cell was investigated for 300 h while operating at a continual density of 100 mA cm2 and generating a cell voltage of 1.0 V. The cell operated effectively for that amount of time, but after 300 h, the stability of the electrode reduced as a result of either mechanical damage or deactivation sustained throughout the operation of the H$_2$-Cl$_2$ fuel cell [25].

5.3 Ruthenium Based Electrocatalysts

One of the most potent electrocatalysts for CER is ruthenium oxide (RuO$_2$), while pure metals are more prone to further oxidation, making them less stable than oxides. Away from its superior catalytic qualities, utilizing RuO$_2$ as an electrode material certifies the saving cost because, regardless of not being cheap, this metal is significantly less expensive than the rest of the Pt-group materials, and it also improves the stability and efficiency of chlorine evolution. RuO_2 is primarily utilized to produce chlorine (MMOs) as a component of mixed metal oxides. Long service lifetimes for the bare oxide electrodes have been established [26]. However, they erode due to the strong reaction environments of the evolution of oxygen and chlorine to which the electrode is exposed, among other things, in manufacturing Chlor-alkali operations. RuO_2 is, as a result, combined with other oxides like IrO_2, Ta_2O_5, TiO_2, and ZrO_2 to enhance coating stability for chlorine evolution and improve corrosion resistance. Ru supplies are running out, which poses an issue even if RuO_2 is less expensive than other materials like platinum. As a result, new electrocatalysts must be created to decrease the amount consumed while improving its electrocatalytic characteristics [27].

RuO_2/TiO_2 was one of the electrode materials used in 1979 to test the electrochemical reduction and oxidation of chlorine and hydrogen. This study demonstrated that the chlorine evolution reaction (CER) had lower activation loss at the electrode on its chlorine side than the oxygen evolution reaction (OER), and it was also faster and reversible. After studying the impacts of the catalyst, they also started a recovering hydrogen-chlorine system using these electrodes, which produced a greatest electric-to-electric efficacy of 95%. In recent experiments, the hydrogen-chlorine system has been found to create energy more successfully. Thomassen et al. investigated the theories behind the revolutionary fuel cells of hydrogen-chlorine in 2006 and came to the conclusion that, when in the chlorine reaction, RuO_2 is utilized as a catalyst, the kinetics of Chlorine reduction is substantially quicker than oxygen reduction reaction (ORR) [28]. Due to the electrode's deterioration, this study did not yield any pertinent data. Recently, various fresh methods have been created to improve and lessen the utilization of Ru as a catalyst.

Similarly, nano casting is a highly intriguing method that Han et al. employed to create a replica of mesoporous RuO_2 that was structurally regulated, ordered, and mesoporous KIT-6 silica was used as a pattern. According to this study, this novel catalyst improved the electrocatalytic activity of economic RuO_2 nanoparticles. The peak current density of SC RuO_2 was 1.75 times greater for CER than that of RuO_2 nanoparticles.

RuO_2 is unstable in acidic settings; hence it has undergone advancements to be used as an electrocatalyst for the chlorine reaction [29]. In 2020, Goryachev et al. offered RuO_2 (110)/Ru (0001) single-crystal model electrodes to test the stability of RuO_2 (110) in the presence of acidic chlorine and oxygen evolution reaction conditions in 0.5 M HCl and 0.5 M H_2SO_4 electrolytes [29]. Black TiO_2 nanotube arrays containing RuO_2 were created by pulsed electrodeposition by [30], and they were then tested as three-dimensional assisting electrodes for hydrogen and chlorine evolution. With a faradaic efficiency of 95.25 percent, black Titania fabricated with RuO_2 demonstrated strong chlorine evolution activity [30].

However, Titania with RuO_2 loaded did not exhibit activity, although having comparable hydrogen evolution reaction performance. The significance of the Titania support type in showing properties at chlorine evolution activity and anodic potentials was highlighted by this fact. Additionally, it was shown that black Titania materials could boost anodic processes' electrocatalytic activity [31].

5.4   Ru Incorporated Bimetallic Based Electrocatalysts

It should be noted that RuO_2 has the lowest overpotential and is the greatest affordable catalyst for the chlorine evolution reaction. However, its catalytic activity is decreased due to ruthenium's increased oxidation state from prolonged operation [31]. To use effective electrocatalysts in reversible fuel cells of hydrogen-chlorine PEM, two or more metals and metal oxides may be combined to alter the surface composition, morphology, or microstructure of the electrocatalyst components.

Many metal oxides are employed as electrocatalyst materials in the Chlor-alkali industry to enhance CER performance. Additional metal oxides are also used to increase the stability of coating during extended operation at high anodic potentials. Coatings made of RuO_2 have high electrocatalytic properties, such as pure RuO_2 and RuO_2-TiO_2. Because of its stability and catalytic activity, RuO_2 combined with titanium oxide is one of the most commonly used catalysts for chlorine evolution [32]. Various variables, including surface area and composition, influence the ability of RuO_2-TiO_2 electrodes to evolve chlorine. To improve the performance of CER, Luu et al. proposed a brand-new electrocatalyst model in 2017 that is based on TiO_2 and RuO_2 mixed with polystyrene templates [33]. Their research identified a potential electrode material with high efficiency of chlorine evolution, and its application can also result in lower energy costs. The most favorable mechanical characteristics are found in iridium oxides (IrO_2). The performance of the CER can be increased, for instance, because the electrical conductivity is greater than that of commercial catalysts. But compared to the chlorine evolution reaction, IrO_2 is more stable. Due to the beneficial characteristics of both oxides, a mixture of RuO_2 and IrO_2 can be a suitable electrocatalyst to improve the performance of CER in a reversible hydrogen-chlorine system [34]. By using TiO_2 atomic layer deposition, [35] suggested improving the catalytic activity of OER and CER electrocatalysts [35]. They demonstrated that OER performance, as measured by Ezc (potentials of zero charge), had enhanced nine times above that of the traditional iridium catalyst. Other non-noble metal catalysts, like Mo, Ni, and Co, have recently demonstrated catalytic performance in acid media comparable to that of Pt. To give two examples, as compared to commercial RuO_2 electrodes, Ru-Co and Ru-Mn display more catalytic activity, improved electrical conductivity, and better stability, Mondal et al. prepared and evaluated two innovative electrode materials built on Mn, Co, and Ru oxides. The study showed that these electrodes were more durable than pure RuO_2 electrodes, and as a result, new developments in these electrocatalysts for the hydrogen halogen regeneration fuel cell are anticipated. Interesting alternatives are the Ru-Mn and Ru-Co alloys, which are created using wet chemical synthesis techniques and coated on titanium metal substrates. When compared to commercial chloride oxidation electrodes, they have a much lower precious metal composition and display high catalytic capacity, strong electrical conductivity, and good stability (MMO). Ru is highly expensive, as was previously said, thus when it was tested, its content was only 1% and it was alloyed with less expensive metals like cobalt and manganese. Stability tests revealed that the electrochemical activity had been kept up to par, and there had been very little mass loss.

Furthermore, compared to commercially available MMOs or pure RuO_2, the performance of halogen redox catalysts was not compromised. A high-performance regenerative fuel cell of hydrogen-chlorine using a $(Ru_{0.09}Co_{0.91})_3O_4$ alloy coated on carbon as the chlorine electrode was created by [36]. No discernible activation losses were found at 0.15 mg Ru cm2 noble metal loadings on chlorine electrodes. It is important to note that this study performed a cost analysis of the electrocatalyst loading process. With a Ru price of $3700 kg and a power density of 0.5 W cm^2, it was calculated to be only a fraction of the cost of a grid-scale storage system, or roughly $1.11 kW^{-1}, for precious metals on the chlorine electrode [36].

Two new chlorine redox catalysts, $Ti/Ru_{0.5}Ir_{0.5}O_2$ and $Ti/Ru_{0.3}Ti_{0.7}O_2$ were created to be used in this novel system centered on a study of the impact of the composition of chlorine electrodes on the effectiveness of electro-absorption cells of reversible Chlor-alkali in a study conducted by Carvela et al. in 2020. These electrodes were made using the Pechini method and cutting-edge microwave and laser technologies [37].

While the $Ti/Ru_{0.3}Ti_{0.7}O_2$ electrode showed poor electrocatalytic activity for both HER and CER, the $Ti/Ru_{0.5}Ir_{0.5}O_2$ electrode demonstrated good electrocatalytic activity. However, the creation of oxygen is also favored when Ir is used as a catalyst because CER strives with OER and lowers the conversion rate into chlorine. As a result, it can be argued that the mixed states of metal–metal oxide are significant electrocatalysts for regenerative PEM fuel cells (RFCs).

5.5 Ru-Incorporated Trimetallic Based Electrocatalysts

To enhance the selectivity, activity, and stability, catalytic capabilities of these electrodes throughout CER, as mentioned in the preceding sections, the electrocatalysts usage based on the mixture of several metal oxides has been demonstrated to be an operative technique. A distinct approach assembled on the inclusion of an innovative metal oxide has been established in many research projects, paving the way to the valuation of ternary oxide systems in light of the beneficial results produced by linking two metal oxides [38].

Based on studies that integrated the usage of three metal oxides as electrocatalysts, it might be concluded that this method is a realistic choice to explore. The combination of metal and metal oxides may provide an effective catalyst to improve the processes of CER and CRR. Nowadays, the most popular electrodes are composed of a TiO_2-RuO_2-IrO_2 mixture placed on a Ti support. Yi et al. investigated the impact of loading of IrO2on TiO_2-RuO_2-IrO_2 electrodes. They created three electrodes in this study by varying the amounts of IrO_2: 0.3, 0.4, 0.5, 0.9, and 1.0 mgTiO_2 cm^2.

The study established that the lifetime of the electrode having the highest content of IrO_2 was nearly double that of the electrode with the lowest level [39]. According to research done in 2009 by Panic et al., the addition of colloidal iridium oxide to sol–gel process RuO_2-TiO_2 coatings resulted in greater activity for OER and CER as well as a decrease in the corrosion rate of $Ti0.6Ru0.4O_2$ coatings on titanium when compared to coatings made using the traditional thermal decomposition method [40]. The electrocatalytic activity, capacitance, and stability characteristics of titanium electrodes stimulated by RuO_2-TiO_2 coatings were improved by IrO_2. IrO_2 was added, which resulted in less CER activity [41].

5.6 Other Materials Based Electrocatalysts

Other substances, including TiO_2, Ta_2O_5, and ZrO_2, have been considered to increase the strength of these active oxides [42]. To increase the electrochemical electrocatalytic activity and active surface area, [43] examined the surface morphologies of electrodes of IrO_2-Ta_2O_5-TiO_2 with various mole fractions of TiO_2. They found that several fine grains developed on the electrode surfaces. The electrode with the best CER performance is IrO_2-Ta_2O_5-TiO_2, with a loading of 0.7 mg cm2. It has a low Tafel slope (50.1–51.5 mV dec1) and a highly electrochemically active surface. The simultaneous very good chloride evolution selectivity of the IrO_2-Ta_2O_5-TiO_2 electrode has significantly reduced oxygen evolution [43].

Due to the significantly lower potential UOER of 1.23 V vs. HER, which regretfully results in the selectivity issue of Cl_2 gas production at the anode, the evolution of gaseous oxygen is more thermodynamically favored at the anode [21]. As a result, the ideal CER electrocatalyst must associate selectivity with high activity in an acidic electrolyte. For CER, the selectivity of these two monolayers is further examined because it is projected that they will perform the best in Cl_2 production [44].

The four electron transfer phases of the OER are carried out as follows using the CHE method:

$$O^* + H_2O \rightarrow HOO^* + H^+ + e^- \tag{5.6}$$

$$HOO^* \rightarrow {}^* + O_2 + H^+ + e^- \tag{5.7}$$

In the course of the OER process, successive intermediates of O*, HOO*, and HO* are generated after corresponding elementary processes. The selectivity for -phosphorene and -arsenene monolayers was assessed in this work using the free energy absorption of the HO* intermediate formed in the first elementary stage of OER. To identify the more energetically stable structures at pH $= 0$, the Gibbs free energy varies of the Cl* and HO* species of - -arsenene and phosphorene as a role of potential applied USHE were calculated [45].

The value of GHO* for the monolayer of -phosphorene is discovered to be lower than that of GCl*, indicating that the development of the HO* intermediate on the -phosphorene is more advantageous from an energy standpoint in an acidic solution. It demonstrates that at the potential region of UCER, the OER process is favored on the -phosphorene monolayer, resulting in poor Cl_2 selectivity [46]. The fact that the value of GCl* for a monolayer of -arsenene is significantly less than that of GHO* indicates that the CER process occurs before the OER has the Cl* precursor, which is more thermodynamically advantageous [47].

References

1. Wang Y et al (2021) Recent advances in electrocatalytic chloride oxidation for chlorine gas production. J Mater Chem A 9:18974–18993
2. Hansen HA et al (2010) Electrochemical chlorine evolution at rutile oxide (110) surfaces. Phys Chem Chem Phys 12(1):283–290
3. Hepel T, Pollak FH, O'Grady WE (1986) Chlorine evolution and reduction processes at oriented single-crystal RuO_2 electrodes. J Electrochem Soc 133(1):69
4. Exner KS et al (2014) Chlorine evolution reaction on RuO_2 (110): Ab initio atomistic thermodynamics study-pourbaix diagrams. Electrochim Acta 120:460–466
5. Chen Y et al (2022) Tip-intensified interfacial microenvironment reconstruction promotes an electrocatalytic chlorine evolution reaction. ACS Catalysis 12(22): 14376–14386
6. Liu J et al (2022) TMN4 complex embedded graphene as efficient and selective electrocatalysts for chlorine evolution reactions. J Electroanal Chem 907:116071
7. Lim T et al (2020) Atomically dispersed Pt–N4 sites as efficient and selective electrocatalysts for the chlorine evolution reaction. Nat Commun 11(1):412
8. Lim, T., et al., General efficacy of atomically dispersed Pt catalysts for the chlorine evolution reaction: potential-dependent switching of the kinetics and mechanism. ACS Catalysis 11(19):12232–12246
9. Liang N-N et al (2024) Selective chlorine evolution reaction using Pt-RuO_2 electrocatalysts in brackish water and seawater at circum-neutral pH and its application to seawater desalination. J Environ Chem Eng 12(5):113622
10. Sun H et al (2024) Unlocking the catalytic potential of platinum single atoms for industry-level current density chlorine tolerance hydrogen generation. Adv Funct Mater 2408872
11. Jadhav AR et al (2024) Stable and Efficient Chlorine Evolution Reaction with Atomically Dispersed Ru on Surface Tensile Strained TiO_2. Appl Catal B 359:124456
12. Gong H et al (2024) Ru dopant induced high selectivity and stability of ternary RuSnTi electrode toward chlorine evolution reaction. Appl Catal B 349:123892
13. Choi WI et al (2023) Ru-Doped Co3O4 nanoparticles as efficient and stable electrocatalysts for the chlorine evolution reaction. ACS Omega 8(38):35034–35043
14. Chen J et al (2022) Atomic ruthenium coordinated with chlorine and nitrogen as efficient and multifunctional electrocatalyst for overall water splitting and rechargeable zinc-air battery. Chem Eng J 441:136078
15. Dang Y et al (2023) Enhanced alkaline/seawater hydrogen evolution reaction performance of NiSe2 by ruthenium and tungsten bimetal doping. Int J Hydrogen Energy 48(45):17035–17044
16. Zhang D et al (2024) Engineering antibonding orbital occupancy of $NiMoO_4$-supported Ru nanoparticles for enhanced chlorine evolution reaction. J Colloid Interface Sci 672:423–430
17. Islam M et al (2023) Bimetallic atom dual-doped MoS_2-based heterostructures as a high-efficiency catalyst to boost solar-assisted alkaline seawater electrolysis. ACS Sustain Chem Eng 11(17):6688–6697

18. Zhou L et al (2023) Bimetallic substrate induction synthesis of binder-free electrocatalysts for stable seawater oxidation at industrial current densities. Chem Eng J 458: 141457
19. Jin M et al (2022) Strategies for designing high-performance hydrogen evolution reaction electrocatalysts at large current densities above 1000 mA cm^{-2}. ACS Nano 16(8):11577–11597
20. Li L et al (2021) Kilogram-scale synthesis and functionalization of carbon dots for superior electrochemical potassium storage. ACS Nano 15(4):6872–6885
21. Carvela M et al (2020) Recent progress in catalysts for hydrogen-chlorine regenerative fuel cells. Catalysts 10(11):1263
22. Chang F Wick H (1935) Untersuchung über die Halogenüberspannung. Zeitschrift für Physikalische Chemie 172(1):448–458
23. Yeo R et al (1980) An electrochemically regenerative hydrogen-chlorine energy storage system: electrode kinetics and cell performance. J Appl Electrochem 10(3):393–404
24. Thomassen M et al (2003) H2/Cl2 fuel cell for co-generation of electricity and HCl. J Appl Electrochem 33(1):9–13
25. Lim T et al (2020) Atomically dispersed Pt–N4 sites as efficient and selective electrocatalysts for the chlorine evolution reaction. Nat Commun 11(1):1–11
26. Iwakura C, Hirao K, Tamura H (1977) Anodic evolution of oxygen on ruthenium in acidic solutions. Electrochim Acta 22(4):329–334
27. Le Luu T soo Kim C Yoon J (2017) Sono-electro-deposition of RuO2 electrodes for high chlorine evolution efficiencies. J Korean Soc Water Wastewater 31(5):397–407
28. Zafar MS et al (2016) New development of anodic electro-catalyst for chlor-alkali industry. Port Electrochim Acta 34(4):257–266
29. Goryachev A et al (2020) Electrochemical stability of RuO2 (110)/Ru (0001) model electrodes in the oxygen and chlorine evolution reactions. Electrochim Acta 336:135713
30. Heo SE et al (2020) Anomalous potential dependence of conducting property in black titania nanotube arrays for electrocatalytic chlorine evolution. J Catal 381:462–467
31. Audichon T et al (2014) Electroactivity of RuO2–IrO2 mixed nanocatalysts toward the oxygen evolution reaction in a water electrolyzer supplied by a solar profile. Int J hydrogen energy 39(30):16785–16796
32. Moradi F, Dehghanian C (2014) Addition of IrO2 to RuO^{2+} TiO2 coated anodes and its effect on electrochemical performance of anodes in acid media. Progress in Natural Science: Materials International 24(2):134–141
33. Tran LL et al (2017) Fabricating macroporous RuO2-TiO2 electrodes using polystyrene templates for high chlorine evolution efficiencies. Desalin Water Treat 77:94–104
34. Kong F-D et al (2012) Pt/porous-IrO2 nanocomposite as promising electrocatalyst for unitized regenerative fuel cell. Electrochem Commun 14(1):63–66
35. Finke CE et al (2019) Enhancing the activity of oxygen-evolution and chlorine-evolution electrocatalysts by atomic layer deposition of TiO 2. Energy Environ Sci 12(1):358–365
36. Huskinson B et al (2012) A high power density, high efficiency hydrogen–chlorine regenerative fuel cell with a low precious metal content catalyst. Energy Environ Sci 5(9):8690–8698
37. Carvela M et al (2020) Effect of the anode composition on the performance of reversible chlor-alkali electro-absorption cells. Sep Purif Technol 248:117017
38. Zeradjanin AR et al (2014) On the faradaic selectivity and the role of surface inhomogeneity during the chlorine evolution reaction on ternary Ti–Ru–Ir mixed metal oxide electrocatalysts. Phys Chem Chem Phys 16(27):13741–13747
39. Yi Z et al (2007) Effect of IrO2 loading on RuO2–IrO2–TiO2 anodes: A study of microstructure and working life for the chlorine evolution reaction. Ceram Int 33(6):1087–1091
40. Panić VV et al (2010) The effect of the addition of colloidal iridium oxide into sol–gel obtained titanium and ruthenium oxide coatings on titanium on their electrochemical properties. Phys Chem Chem Phys 12(27):7521–7528
41. Weibel A, Bouchet R, Knauth P (2006) Electrical properties and defect chemistry of anatase (TiO2). Solid State Ionics 177(3–4):229–236
42. Oakton E et al (2017) IrO2-TiO2: a high-surface-area, active, and stable electrocatalyst for the oxygen evolution reaction. ACS Catal 7(4):2346–2352

43. Deng L et al (2019) Preparation of electrolyzed oxidizing water by TiO_2 doped IrO_2-Ta_2O_5 electrode with high selectivity and stability for chlorine evolution. J Electroanal Chem 832:459–466

44. Saha S, Kishor K, Pala RG (2018) Modulating selectivity in CER and OER through doped RuO_2. ECS Trans 85(12):201

45. Liu J et al (2022) Low-Dimensional Metal-Organic Frameworks with High Activity and Selectivity toward Electrocatalytic Chlorine Evolution Reactions. The Journal of Physical Chemistry C 126(16):7066–7075

46. Khatun S, Hirani H, Roy P (2021) Seawater electrocatalysis: activity and selectivity. J Mater Chem A 9(1):74–86

47. Lim T et al (2019) Promoting Activity and Selectivity of Electrochemical Chlorine Evolution Reaction by Atomically Dispersed Pt Catalysts. Korean Chem Soc

Chapter 6
Conclusion and Future Perspectives

Seawater is generally seen as an infinite resource when compared to pure water. Since seawater is an inexhaustible resource, producing hydrogen via electrolysis can help to some extent with the world's energy issue. Seawater electrolysis still has a lot of room for development for the high-performance production of hydrogen despite the numerous efforts put into the field in recent years. The presence of numerous cations and anions in seawater makes it clear that it is more complicated than freshwater electrolysis. Although direct seawater splitting is affordable and environmentally beneficial, its practical implementation is seriously affected by the fundamental difficulties in seawater, including the challenging processes associated with oxygen evolution reaction (OER), chlorine evolution reaction (CER), and hydrogen evolution reaction (HER). A promising method is the production of electrocatalysts found in abundance on Earth for direct seawater splitting. To avoid the influence of side reactions and resist diverse impurities, highly effective electrocatalysts with better selectivity and stability are important. The first section of this review covers the understanding of direct seawater electrolysis, while the next sections focus on its mechanism and different types of catalysts involved in seawater splitting. We have discussed a variety of materials used as catalysts, with noble metal catalysts, metal nitrides, metal borides, non-noble metal catalysts, metal phosphides, metal oxides, and hydroxide), metal sulfides and hybrid metal catalysts for seawater splitting. The different types of electrocatalysis used in OER, HER, and CER are discussed in Tables 1, 2, and 3, respectively, with their reported catalytic performance. We conclude that this review will help in research on seawater splitting in the future.

After a though study of the seawater electrolysis research, we think there is a need to explore and develop new electrolysis with high stability and activity. It should be noted that numerous cations and chloride anions in seawater inhibit water-splitting processes. The synthesis of catalysts with better HER, OER, and CER selectivity than other competing reactions is greatly desired. Modifying the electrical structure of active sites is extremely important and significantly impacts catalytic performance to produce highly effective electrocatalysts for seawater electrolysis. Moreover, the

Z. K. Ghouri et al., *Hydrogen Production from Seawater Electrolysis*,
SpringerBriefs in Energy, https://doi.org/10.1007/978-3-031-73442-7_6

Catalysts are the main focus of the current seawater electrolysis research for the creation of hydrogen. Instead of focusing solely on the catalysts, the complete reactor must be considered to effectuate the electrocatalytic production of hydrogen. Reactors that can be modified for particular seawater electrolysis must be thoughtfully designed. A perfect anode for seawater electrolysis must include effective electron and mass transfer properties, selective exposed active capacities with high mechanical and structural stability, corrosion resistance, and intrinsic activity. Despite considerable advancements and current research initiatives, direct seawater electrolysis still has a long way to go before it can produce green hydrogen that is both economical and sustainable. Some perspectives on using enhanced anode materials and electrocatalytic techniques in future research are mentioned.